PURCHASING AND QUALITY

QUALITY AND RELIABILITY

A Series Edited by

Edward G. Schilling
Center for Quality and Applied Statistics
Rochester Institute of Technology
Rochester, New York

1. Designing for Minimal Maintenance Expense: The Practical Application of Reliability and Maintainability, *Marvin A. Moss*
2. Quality Control for Profit, Second Edition, Revised and Expanded, *Ronald H. Lester, Norbert L. Enrick, and Harry E. Mottley, Jr.*
3. QCPAC: Statistical Quality Control on the IBM PC, *Steven M. Zimmerman and Leo M. Conrad*
4. Quality by Experimental Design, *Thomas B. Barker*
5. Applications of Quality Control in the Service Industry, *A. C. Rosander*
6. Integrated Product Testing and Evaluating: A Systems Approach to Improve Reliability and Quality, Revised Edition, *Harold L. Gilmore and Herbert C. Schwartz*
7. Quality Management Handbook, *edited by Loren Walsh, Ralph Wurster, and Raymond J. Kimber*
8. Statistical Process Control: A Guide for Implementation, *Roger W. Berger and Thomas Hart*
9. Quality Circles: Selected Readings, *edited by Roger W. Berger and David L. Shores*
10. Quality and Productivity for Bankers and Financial Managers, *William J. Latzko*
11. Poor-Quality Cost, *H. James Harrington*
12. Human Resources Management, *edited by Jill P. Kern, John J. Riley, and Louis N. Jones*
13. The Good and the Bad News About Quality, *Edward M. Schrock and Henry L. Lefevre*
14. Engineering Design for Producibility and Reliability, *John W. Priest*
15. Statistical Process Control in Automated Manufacturing, *J. Bert Keats and Norma Faris Hubele*

16. Automated Inspection and Quality Assurance, *Stanley L. Robinson and Richard K. Miller*
17. Defect Prevention: Use of Simple Statistical Tools, *Victor E. Kane*
18. Defect Prevention: Use of Simple Statistical Tools, Solutions Manual, *Victor E. Kane*
19. Purchasing and Quality, *Max McRobb*
20. Specification Writing and Management, *Max McRobb*

PURCHASING AND QUALITY

Max McRobb
Bedford, England

MARCEL DEKKER, INC.   New York and Basel

Library of Congress Cataloging-in-Publication Data
McRobb, Max
  Purchasing and quality / Max McRobb.
    p. cm.
  Includes index.
  ISBN 0-8247-8075-2 (alk. paper)
  1. Quality control. I. Title.
TS156.M389 1989
620'.0045—dc20   89-1427
                                                       CIP

This book is printed on acid-free paper.

Copyright © 1989 by MARCEL DEKKER, INC. All Rights Reserved

Neither this book nor any part may be reproduced or transmitted in any form or by any means, electronic or mechanical, including photocopying, microfilming, and recording, or by any information storage and retrieval system, without permission in writing from the publisher.

MARCEL DEKKER, INC.
270 Madison Avenue, New York, New York 10016

Current printing (last digit):
10 9 8 7 6 5 4 3 2 1

PRINTED IN THE UNITED STATES OF AMERICA

About the Series

The genesis of modern methods of quality and reliability will be found in a simple memo dated May 16, 1924, in which Walter A. Shewhart proposed the control chart for the analysis of inspection data. This led to a broadening of the concept of inspection from emphasis on detection and correction of defective material to control of quality through analysis and prevention of quality problems. Subsequent concern for product performance in the hands of the user stimulated development of the systems and techniques of reliability. Emphasis on the consumer as the ultimate judge of quality serves as the catalyst to bring about the integration of the methodology of quality with that of reliability. Thus, the innovations that came out of the control chart spawned a philosophy of control of quality and reliability that has come to include not only the methodology of the statistical sciences and engineering, but also the use of appropriate management methods together with various motivational procedures in a concerted effort dedicated to quality improvement.

This series is intended to provide a vehicle to foster interaction of the elements of the modern approach to quality, including statistical applications, quality and reliability engineering, management, and motivational aspects. It is a forum in which the subject matter of these various areas can be brought together to allow for effective integration of appropriate techniques. This will promote the true benefit of each, which can be achieved only through their interaction. In this sense, the whole of quality and reliability is greater than the sum of its parts, as each element augments the others.

The contributors to this series have been encouraged to discuss fundamental concepts as well as methodology, technology, and procedures at the leading edge of the discipline. Thus, new concepts are placed in proper perspective in these evolving disciplines. The series is intended for those in manufacturing, engineering, and marketing and management, as well as the consuming public, all of whom have an interest and stake in the improvement and maintenance of quality and reliability in the products and services that are the lifeblood of the economic system.

The modern approach to quality and reliability concerns excellence: excellence when the product is designed, excellence when the product is made, excellence as the product is used, and excellence throughout its lifetime. But excellence does not result without effort, and products and services of superior quality and reliability require an appropriate combination of statistical, engineering, management, and motivational effort. This effort can be directed for maximum benefit only in light of timely knowledge of approaches and methods that have been developed and are available in these areas of expertise. Within the volumes of this series, the reader will find the means to create, control, correct, and improve quality and reliability in ways that are cost effective, that enhance productivity, and that create a motivational atmosphere that is harmonious and constructive. It is dedicated to that end and to the readers whose study of quality and reliability will lead to greater understanding of their products, their processes, their workplaces, and themselves.

Edward G. Schilling

Preface

This book has been designed primarily for those who are new, or comparatively so, to the buying profession. However, the range of subjects and techniques and the detail of attention given should prove of value even to experienced buyers. When the manuscript, in summary form, was read by a senior buyer with many years of experience in a large, high-tech company, he was enthusiastic that such a book was being made available.

This book will also be of value to all those engaged in the "quality scene" who may have any involvement with bought-out quality audit (BOQA), vendor quality assurance (VQA), or any related activities.

This book has been built up from many years of experience, and it should be emphasized that all of the situations described did really happen and that I participated in most, if not all, of them. The reader will find that there are appendixes to Chapters 1, 3, 5, 6, 10, 12, and 13. Most of these appendixes are intended to provide basic, factual data for the reader who wishes to delve in more detail into the real solutions that were developed in particular cases and, perhaps, adapt them to suit his own particular needs.

The reader will find that 14 of the 19 appendixes are actual operational procedures from the quality manuals of three companies identified as "A", "B", and "C". Company A is a medium-size precision engineering company in a very specialized field, with several hundred employees, a high proportion of its output going to the Armed Forces of the United Kingdom. Company B is quite the reverse. It has about 24 employees and its customers are mainly very large industrial continuous-process companies. It designs its own products using only standard bought-out items, which are then assembled "on site." One has in these two companies a complete contrast in organizational requirements displayed with such different sizes and product complexity. Company D is a leading light engineering organization in the consumer products field, with several thousand employees. Thus, three principal industrial fields are covered: military, general industrial, and consumer products. These companies also illustrate the wide range of industrial scenarios to which the principles discussed in this book can be applied. Company C is featured in case studies in Chapter 8 and is a large internationally known company. Its product is medium-heavy precision engineering and is a world leader in its field.

The sizes of the companies whose purchasing and quality arrangements have formed the basis for this book range from very large to very small. It might be thought that procedures devised for large companies would be of little use to very small companies. This is not the case. In fact, all the procedures that were written for Company B were based on ones originally written for a very large company (similar, in fact, to companies A and D).

The operational procedures that were written for company B were also based on the requirements of British Standard (BS) 5750, a standard for quality systems, that has often been called unsuitable for small companies. This, obviously, is not the case, but it is necessary to exercise some discretion when applying the requirements of BS 5750 to the organizational arrangements of a small company. This comment applies with equal force whether one is considering BS 5750, the British Military Standards in the series 05-21 to 05-29, the American MIL-Q-9858A, the North Atlantic Treaty Organization (NATO) AQAP series, or the International Standards Organization (ISO) 9000 series. They are all basically the same. (But see Chapter 5 and its Appendix 5.)

The sampling inspection guide, chapter 5 Appendix 1, was designed by me for Company C, admittedly many years ago. The reason for including it here is that my more recent experience has shown that such a guide is still very necessary. Both large and

PREFACE

small companies often do not understand the significance or the proper operation of inspection sampling plans to, for example, MIL-S-105D.

Appendixes 4 and 5 of Chapter 5 are also worthy of mention here. Appendix 4 contains the full text of the NATO AQAP 4. This details the requirements for the middle level of the usual three levels of quality system requirements found in the various standards of this kind. Appendix 5 is the full text of a self-determination program that I devised. It is intended to enable companies to "pre-assess" the probable degree of conformance to any one of the three levels of requirements in BS 5750, that is to say, before an official assessment on behalf of a customer. It is also easily adaptable to any of the other national, or international, standards that are referred to throughout the book.

There are a number of acknowledgments I must make. First, to the editors of the British journals *Purchasing* and *Purchasing Management* and the editor of the American journal *Quality*, for their permission to reprint material taken from some of the numerous articles written for those journals or their predecessors; to Mr. D. G. Spickernell, (then) Director General of the British Standards Institution, for permission to print extracts from BS 5750; to the Institute of Quality Assurance for permission to include the requirements for Registered Lead Assessors and the Code of Practice which applies to those Assessors; to the World Headquarters of ITT for their permission to print details of their Quality Cost Improvement Plan; to the British Ministry of Defence for supplying the copies of the various AQAPs, and permitting the reprinting of the various extracts from them and of AQAP 4 in full. The last formal acknowledgment is to Her Majesty's Stationery Office in London for permission to reproduce the extracts from the National Economic Development Council Report that appear in Chapter 1.

Finally, I feel that I really must dedicate this book to my wife, Cath, for her patience in putting up with a considerable amount of domestic dislocation during the many months it took to write this book, to say nothing of acting as a proofreader! A word processor, printer, and piles of paper scattered around the living room are not exactly conducive to the fine art of household management.

Max McRobb

Contents

About the Series   *iii*
Preface   *v*
Contents   *ix*

1	**The Changing Role of the Buyer**	1
	Appendix: Purchasing Responsibilities in the Quality System	9
2	**Specifications for Buyers**	13
3	**What Is Being Ordered**	23
	Appendix: Design Responsibilities in the Quality System	27
4	**The Order**	37
5	**Vendor Selection and Approval**	43
	Appendix 1: A Simple Guide to Sampling Inspection	55
	Appendix 2: Assessment, Selection, and Control of Subcontractors	59
	Appendix 3: Requirements for Subcontractor Gauge Control and Calibration	69

	Appendix 4: NATO Inspection System Requirements for Industry	71
	Appendix 5: A Self-Determination Program for the Estimation of the Degree of Conformity to the Three System Levels in Parts 1, 2, and 3 of BS 5750/1987 (or ISO9001/3)	78
6	**When It Is Received**	**93**
	Appendix 1: General Procedure Governing Incoming Inspection	105
	Appendix 2: Metallurgical and Chemical Inspection of Incoming Material	111
	Appendix 3: Control of Incoming Material	113
7	**Vendor Rating and Surveillance**	**117**
8	**CAST (Customer and Suppliers Together)**	**127**
9	**The Why and Wherefore of Vendor Quality Assurance**	**137**
10	**The Importance of Specifications to BOQA and VQA**	**147**
	Appendix 1: Procedure for Quarantine and Bonded Storage	152
	Appendix 2: Procedure for Supplier Initial Production Sample Approval	154
11	**Vendor Quality Assurance, Liaison and Backup**	**159**
12	**Selecting and Training Surveyors/Assessors**	**167**
	Appendix 1: Extracts from the "Rules and Procedures of the Lead Assessor Certification Scheme"	171
13	**Record Systems**	**175**
	Appendix 1: Procedure for Concessions and Production Permits	185
	Appendix 2: Procedure for Requirements for Subcontractor Records	189
	Appendix 3: Procedure for Inspection Documentation and Records for Goods Inwards Inspection	191

CONTENTS

	Appendix 4: Procedure for Purchasing Documentation	194
	Appendix 5: Procedure for Retention and Disposal of Records	195
14	**Quality Costs**	**199**
Appendix		*211*
Index		*215*

PURCHASING AND QUALITY

1
The Changing Role of the Buyer

Some years ago some very scathing criticisms of buyers in general appeared in a respected British purchasing journal. Of the many comments made, the following were the most pointed:

"There are quite a few times when I've thought that 90% of the purchasing profession have their feet screwed to the office floor!"

"The sheer damned inertia of buyers... drowned... in documentation routines."

"Most... don't realise the enormous scope that they have in the business sense."

"No opportunity should be allowed to pass... and his stature will have been improved enormously...."

These comments were attributed to an anonymous purchasing director who was presumably in a position to know.

I can clearly remember the buyer of the engineering company in the 1930s, in which I served my apprenticeship, and his general methods of operation. When one has seen the methods of operation of a good many of today's buyers, it is hard to realize that their methods are not so very different in many cases from those used by that buyer a half-century ago. Yet in the intervening period there have been many new ideas, many new techniques, to add to the armory of the buyers to increase their scope and authority. Whether or not that purchasing director was right, there can be no doubt at all that today's buyers must keep in the forefront of developments if they are to progress and succeed. But there are some who do not do this.

For a good many years I have been in a somewhat priveleged position which has given me an overview of the work of many buyers at all levels of the profession. Indeed, for more than a decade I have lectured to many hundreds of buyers in purchasing courses. Not surprisingly, my subject was "Quality in Purchasing." Many were keen and eager to learn as much as possible about their professional work, but it was also apparent that there was often not enough appreciation of the ever-widening scope of the business activities that were being developed and which lay ahead of them. Equally, it is the case that there are many in the profession who have advanced with the times, taken up the new ideas, and trained others in them. But there have not been enough such people. It is not clear why this should be so. Perhaps the juniors are not exerting enough "push" on their seniors? Perhaps the seniors are not providing enough guidance and backing to the development of the juniors?

However, buyers should realize that today, in the latter part of the twentieth century, some of the most exciting developments that the profession has yet seen are taking place. They are developments that will provide a rich reward for buyers who are prepared to invest in them by learning, and who are also prepared to show that they can, and do, take advantage of these opportunities to adopt new activities which are so different from the old, traditional buying image. That buyer of the 1930s would indeed be surprised if he could come back today and see what today's buyers do routinely.

Larger buying departments have set up specialist sections in new buying activities so that they may take advantage of the benefits that will accrue to them as a result. The work that these specialist sections undertake is unfamiliar to regular buyers but is full of its own interest and value to the organization. It enables the buying department to function much more effectively in today's increasingly complex industrial situations. There are a number of

CHANGING ROLE OF THE BUYER 3

these new activities of which the following are among the better known:

Engineering
Resource planning
Research
Procurement

There is, however, one additional activity, which is the principal foundation stone for this book:

QUALITY

Some discussion about the ways in which the first four of these new activities are taking their place on the buying scene will help to prepare for the general progression toward QUALITY which is to be followed in this book.

When one mentions an engineering section in a buying department, the immediate reaction will probably be to think of "engineering" buyers, who specialize in buying all kinds of engineering product—in much the same way that in the buying departments of large chain stores one will find specialist buyers for shirts, or in a large engineering company, a specialist gauge buyer. But this is not what this kind of engineering section does at all. It does not buy. It must, of course, be familiar with the buying programs of the department as a whole so that it can develop to the full the particular specialist advice and guidance to buyers that it has been set up to provide for them. It will advise on new materials and methods that may be used as substitutes. Evaluations will also be made of the engineering capabilities and prospects of suppliers so that the "buying" buyers can have all the data they need to help them arrive at their "best" decisions when selecting suppliers.

Resource planning is an activity in which interest was boosted by the oil crisis of the early 1970s, which caused shortages in very many commodities. It may also be regarded as being rather exotic. It deals largely with the future use of all available resources, ranging from manpower through materials to energy. It also concerns itself with the acquisition, storage, distribution, and working properties of those resources. It is concerned with reducing scrap and wastage in increasingly scarce materials and in reducing the scale of environmental pollution.

Resource planners provide buyers with information and guidance about the maximization of the economic results of their work. They will help the buyers in decision-making processes when consideration has to be given to material substitution problems arising because of shortages or lengthening lead times. This also supplements the engineering guidance available. A great deal of the time of resource planners is spent in investigations to determine future patterns in the use of resources. This work produces benefits for other parts of the organization as well. For example, designers can be alerted to possible future new materials in time to ensure incorporation in new designs and to overcome the problems of material runout.

Allied to resource planning and, to a certain extent, with overlapping activities, is purchasing research. It is somewhat more concerned with general forecasting and product planning. It will deal with economic indicators, market behavior, forward buying problems, forecasting techniques, and the development of statistical models. One of the more interesting developments in this field was the introduction of life-cycle cost models for procurement purposes. A leading purpose for using such a model is to take account of the concept of "failure costs."

Considering only the costs of a project, the model is an element in the analysis of the overall benefit. A customer will determine the overall life-cycle cost that he is prepared to accept and the proportions which will be available for the initial investment and the failure costs. There are a number of additional factors which enter into supplier proposed costs and prospective suppliers may offer a combination of them which best suits their own particular circumstances. From this information, provided by the suppliers, the customer is able to consider the various cost elements which will arise during the entire life of the project. It is then possible for the customer to select from those offered, the most suitable combination that most nearly suits the circumstances.

Vaguely associated with research are activities that can be lumped together under the heading "procurement." There has been, and no doubt will continue to be, arguments about the meaning of the word "procurement" in this context. However, one can safely say that it is concerned with all aspects of a project, from conception to eventual obsolescence. "From cradle to grave" is a simple way of putting it. Because of the breadth of its scope, it could be called the *total life* concept. But it also encompasses the quality activity.

Perhaps the biggest and best example of appreciation of the significance of the importance of "procurement" was the creation some years ago of the Procurement Executive of the British Ministry of Defence. This new organization took over responsibility for all activities previously undertaken by various branches of the ministry: activities such as development, contracts, quality, and others. It followed from a very detailed investigation of the work of the ministry made by a senior director of one of the largest and most successful chain stores in the United Kingdom, who was "lent" to the government by his company for more than a year to carry out the task. The principal outcomes of this major reorganization were the significantly improved efficiency of defense procurement, and lower costs.

With the possible exception of the life-cycle cost model, it is likely that many buyers would agree that the various activities which have been discussed so far would fall into the scope of activities in which progressive buyers would expect to have some degree of involvement. *But what about quality?*

Not only in the field of quality itself, but in most other industrial disciplines as well, it was fashionable to regard all quality tasks as being the sole responsibility of the inspection and/or quality departments. This is still the case in many companies, as the validity of the premise that all departments have a part to play in the achievement of quality is slow to gain recognition. At the very least, little regard has been given to the thesis that other industrial disciplines, including buying, have some involvement in the achievement of quality. Sometimes the degree of involvement can be considerable, with an equally large influence on the outcome—in some cases even more than that of the inspection/quality departments themselves.

Buyers may have some difficulty in accepting this point, even although it is quite possible that they may, in some instances, have more influence, although indirect, than that of the person directly responsible for quality. Before any buyers leap up to protest this statement, let them stop to consider this question: How often, when arriving at a decision about which supplier should get a particular order, is the principal factor under consideration that of the lowest cost? And this always affects quality.

At this point it would be pertinent to reflect on the thoughts of Colonel John Glenn, the U.S. astronaut. In answer to a question as to what his thoughts were in the final few seconds before blast-off, he said: "I thought 'What am I doing here sitting on top of thousands of components all supplied by the lowest bidder?'"

The idea of direct involvement in the achievement of product quality is, for many buyers, not easy to accept because of the traditional, long-held historical attitude toward quality on the part of most industrial disciplines. Nevertheless, it is important that all buyers realize and accept this fact of industrial life; that with respect to the material they buy, they have a heavy responsibility for its standard of quality. But it must be said that there are many buyers who are well aware of this fact and act on it, even if it is equally true that there are many buyers who do not (see the extracts at the end of this chapter from the British government's Warner Report). Perhaps one of the first steps such buying departments ought to take is to ensure that there are friendly and close relationships between themselves and the quality department, whatever title that department may have. Long and wide experience has shown that in far too many cases any relationship that exists is at best minimal and at worst may be actively hostile, although there are many with close and friendly relationships. The benefits of such a policy are very great indeed for both the buying department and the company itself.

A major activity in many buying departments is that of cost reduction. But it is usually an independent activity and often undertaken in the literal sense of seeking the lowest first cost, with little regard to product quality. However, a cost-reduction program operated jointly with the quality department can be very rewarding for the company as well as the buying department itself—and the individual buyer.

Consider the case of ITT, one of the world's largest and most successful companies, which operates in a number of very different industries and in many countries. In the late 1960s the company determined that its total "cost of quality" was equal to about 10% of the value of its worldwide sales. This, it was decided, was too much, so the company prepared and introduced a carefully worked out overall quality plan which involved all departments in the company, not the least of which was the buying department. By five years later the results were quite startling. In the words of the president of the company: "The profits for the financial year just ended were $353 millions more than they would have been but for the introduction of that co-ordinated quality cost improvement plan."

Since that time ITT has developed the plan considerably and

CHANGING ROLE OF THE BUYER

published a practical management guide about it. The plan consists, essentially, of 12 steps:*

1. Establish management commitment.
2. Form quality improvement team.
3. Begin quality measurement.
4. Begin "cost of quality" evaluation.
5. Begin "quality awareness" campaign.
6. Begin institution of corrective actions.
7. Begin defect prevention audits.
8. Begin zero-defects program planning.
9. Begin supervisor training.
10. Set goals for cost reduction targets.
11. Begin error cause removal program.
12. Begin recognition programs.

Of course, there are some points that must be made immediately. The first is that by no means all of that extra profit is due to quality and cost improvement resulting from buying department activities. The second point is that both the cost of quality and the scope for increasing (quality) profits will vary considerably from company to company. It must also be realized that that vast sum represented 5% of ITT's sales for the financial year in question, so it is not too difficult to arrive at a reasonable estimate of the comparable figure for any other company. ITT still operates these plans very successfully. The subject of LOST COSTS is discussed more fully in Chapter 14.

Let there be no mistake about it; when one is talking about the benefits accruing from quality as a new role in buying, one is talking about substantial increases in company profits—increases that can only be achieved by:

Close collaboration between the buying and quality disciplines
Much more interest being taken in quality problems on the part of the buyer than is often the case
The acceptance by buyers of their share of the responsibility for the quality of the materials they buy whether raw materials or finished products

---

*I wish to acknowledge with thanks permission received from World Headquarters of the ITT Corporation to paraphrase this material from their booklet "Quality Improvement Through Defect Prevention: A Practical Guide for Management." See also Chapter 14.

There is a variety of quality-related actions that require the involvement of buyers and they are discussed in the appropriate places in subsequent chapters. However, at this point I want to introduce an important document. In May 1985, to provide support for the British government's National Quality Campaign, a task force under the chairmanship of Sir Frederick Warner (current president of the British Institute of Quality Assurance) presented a report to the National Economic Development Council (NEDC), which had appointed it. The NEDC, whose chairman is the chancellor of the exchequer, is made up of representatives of government, management, and unions. It is intended to provide a forum for economic consultations among those three groups.

The report, which bears the title "Quality and Value for Money," is of considerable importance to all sectors of industry. It identifies a number of industrial disciplines as having a particular part to play, one of them being the buying profession. There are three pages in the report that are of special significance for buyers: pages 20, 51, and 55.

On page 20 begins Section 4.5, "Quality Assurance—Procurement and Purchasing." The first sentence in this section is perhaps the most significant. It reads: "The role of buyers in industry in purchasing raw materials, components, sub-contract work etc. is *crucial* [writer's italics] to the achievement of quality since poor quality materials cannot easily be manufactured into high quality goods." Later in the section it says: "This may involve purchase contract clauses which include standards and certification and in some cases involve longer term relationships with suppliers. A number of purchasing organizations have now developed this approach with their suppliers aimed at reducing long term costs. The use of relevant standards, certification schemes and firms of assessed capability by purchasers is often in their own interests since the value of these systems increases as their use spreads."

Two other extracts from this section are of very special relevance to the purpose of this book. The first is: "The buyer needs clear specifications with which he can place orders . . . " (see Chapter 2). The second is: "The buyer also requires feedback on quality performance so that vendors and potential suppliers can be appraised according to their quality performance" (see Chapters 5, 6, and 7).

Section 8.1 of the report beginning on page 51, discusses the role of large buying organizations—in particular, the way in which, in so many cases, they have developed their own in-house systems, which do not bear much similarity to those developed by other organizations. The report also says that differing systems are even developed in different divisions of the same organization. "The potential to improve the situation . . . " is clear. The last sentence in this section is: "The Task Force believes that there is still a great deal of scope for industrial users, collectively, to help the situation and recommends that leading firms in key sectors of industry should be encouraged during the next stage of the Quality Campaign to assist this process."

Finally, note should be taken of Section 8.4 beginning on page 55, "The Role of Professional Institutions," in which the last paragraph says: "The Task Force considers that professional institutions have an important role in promoting good practice and in generating guidance on the way in which quality and standards techniques should be applied within their area of concern. In particular the professional institutions have an important role in translating the general concepts and systems into specific guidance relevant to their interests." An actual procedure that sets out "Purchasing Responsibilities in the Quality System" has been included as an appendix to this chapter. It was prepared for a medium-sized engineering company, for the purposes of this book called Company A, producing complex products. It shows for that particular company just what the purchasing responsibilities were considered to be. A few minor changes and deletions have been made so that the identity of the company is not revealed.

APPENDIX: PURCHASING RESPONSIBILITIES IN THE QUALITY SYSTEM

1. Purpose of the Procedure. To set out the responsibilities of the purchasing department in the overall assurance of quality requirements in the company products.

2. Scope. This procedure covers the activities of the vendor manager and also of the separate purchasing departments in the two divisions of the Company.

3. **Related and Relevant Procedures or Other Documents**

Def. Stan. 05-26, paragraph 211
Def. Stan. 05-21, paragraph 210a, b, c
01-101-003 Documentation Change Control
03-701-001 Requirements for Vendor Records
03-702-001 Assessment, Selection, and Control of Vendors
03-702-002 Evaluating a Vendor
03-711-001 Requirements for Vendor Gauge Control and Calibration
03-713-001 Release of Company Materiel to a Vendor
03-718-002 Purchasing Documentation

4. **Responsibilities.** The prime responsibilities under the requirements of this procedure are those of the vendor manager and of the two chief buyers of the two divisions of the Company. There are, however, associated responsibilities in the design and quality assurance departments connected with the provision of drawings and specifications and with the approval of vendors.

5. **Requirements of the Procedure**

   5.1. Orders shall only be placed with vendors that have been approved according to the requirements of procedures 03-702-001 and 03-702-002.

   5.1.1. If the need arises to place orders upon unapproved vendors, the buyer concerned must immediately contact the quality assurance vendor control engineer (VCE) so that actions to secure approval of the vendor can be put in motion.

   5.2. The purchase order shall provide a clear description of the materiel which is being ordered, including, as may be relevant, the following items of information:

   5.2.1. The type, style, class, grade, or other precise identification.

   5.2.2. The title, or other positive identification, and applicable issue of any drawing, specification, process requirement, inspection instruction, or other relevant technical information.

   5.3. Copies of the relevant documents, referred to in paragraph 5.2.2, must also be sent to the vendor with the order.

   5.4. The order, and any other documents that accompany it, must also make clear the conditions of inspection which apply and including, as may be necessary, the following:

CHANGING ROLE OF THE BUYER

5.4.1. Instructions regarding any preproduction samples which may be considered to be necessary.

5.4.2. Any special requirements concerning packaging.

5.4.3. Requirement for release notes or test certificates as may be appropriate, or any other similar documents.

5.5. In the event that there is a need to make a change in any of the foregoing requirements contained in paragraphs 5.2.1 to 5.4.3 inclusive, the buyer concerned must ensure that every such change is covered by an amendment to the order. Or, in the case of any other change to the order that is not covered by the above.

5.6. The buyer should ensure that copies of each order are provided to the VCE and the quality assurance representative (QAR)* as may be required by them.

5.7. At the earliest possible moment in the ordering program for any contract, the buyer, through the chief inspector, must consult the QAR to determine what his or her requirements are likely to be regarding any source inspection that he or she may wish to call up, or any other relevant requirement.

5.8. To assist the buyer in fulfilling the requirement in paragraph 5.7 he or she must secure from the planning office an advance list of the items that it is proposed to purchase from vendors.

5.8.1. The buyer should ensure that copies of the list are also supplied to the QAR and also to the VCE to assist them.

5.9. The buyer must ensure that he or she maintains the various purchasing records required according to procedure 03-718-002 and, in particular, maintain a register of all specifications/documents sent to particular vendors together with details of the issue/revision of each document concerned.

---

*The quality assurance representative is the representative of the vendor, in this case the British Ministry of Defence.

2
Specifications for Buyers

Conscientious suppliers do their best to give their customers what they ask for. Regrettably, their efforts are sometimes frustrated by a lack of understanding, which is displayed by more than a few buyers, about the purpose and use of specifications. The result is rejected material and hard words all round. Remember the words of the purchasing director at the beginning of the Chapter 1?

It must be made clear that there are many good buyers. Nevertheless, there are too many of the other kind. An example provides support for these statements. An order was received by a component supplier with a specification that had as one of its requirements the following clause: "The components supplied against this specification shall have a guaranteed failure rate not exceeding 0.0001% per thousand hours." This apparently innocuous requirement meant that to provide this guarantee, the supplier would have had to test well over 2000 components without a single failure for more than 100 years.

To a large extent the cause of such situations stems from the specification explosion that rapidly advancing technologies have brought in their wake during the past few decades. It is more or less true to say that 99% of all technological advances achieved since the beginning of time have taken place in this century, and most of it in the last half. This explosion has resulted in a reduction in the standards of specification writing, for which many reasons may be adduced. Among the more significant of them are the following, which should all be of interest to buyers.

1. The system requirements around which specifications are written are increasingly complex.
2. Many of the writers of specifications are physically separated, to a greater or lesser extent, from the people and areas that will be using them. This tends to reduce the extent of consultation.
3. Production pressures accentuate the very real difficulties that many specification writers face in expressing their requirements in clear, unambiguous, language.
4. Impatience is shown by many specification writers when faced with the needs of "lesser mortals" who are trying to understand what they have written.
5. Many specification writers do not fully realize the technical consequences of their specified requirements (e.g., the 100-year example a few paragraphs back).

In the form of procurement specifications of one type or another, all the technical information that is given to a supplier passes through the hands of a buyer—or should do so. Thus, willy-nilly, the buyer takes a good part of the backwash of protest—and back wash there is. As a result, the quality of product suffers for indefinable reasons. When customer procurement specifications are not understood by the suppliers' personnel, one of two scenarios is likely to result:

1. The requirements that are not completely understood will be completely ignored by suppliers or, they will apply their own interpretations of them. In this case there will probably be at least an even chance that the suppliers will be wrong. As a result, it it is more than likely that costly quality problems will arise (most quality problems turn out to be costly).
2. In a few (probably all too few) cases the supplier's personnel will scrutinize incoming specifications in detail and, when doubts arise (all too often), will take them up with the customer at some expense. In this case the outcome will be better specifications—which ought to have been written in the first place.

SPECIFICATIONS FOR BUYERS

One does not have to dip too deeply into one's experience to find examples of the latter situation. In one case it was estimated that the cost to a supplier of "sorting out" a series of problems that arose in a new range of customer specifications was a good many thousand dollars. The fact that the outcome was a much better set of specifications, without problems, should not be allowed to obscure the fact that those specifications had not bee as well written as they should have been in the first instance. A company that maintains a full monitoring service on incoming customer specifications will spend a lot of money in the process, but it will save itself and its customers a great deal more in the long run. (Incidentally, it is more than likely that the cost of the monitoring service and/or the cost of the quality problems that would otherwise have arisen will not be found in any part of the company's accounting system. Nor will the associated costs be found in the accounts of the customer company) (see Chapter 14).

It is unfortunate that to some extent, industry buyers have contributed to this situation through inaction or misdirected interest or even in many cases, ignorance. The latter refers to a lack of knowledge of the full characteristics of the product being bought. In such cases buyers are themselves applying a severe restriction to the professionalism they are bringing to bear on a task. Such restrictions also cut back on their subsequent ability to obtain the right terms for the right product.

Inaction may stem from the fact that many buyers take the view that the content of specifications is "not for them," as they are far too busy concentrating on the commercial aspects of the purchase. Specifications are the responsibility of the technical department. This appears to be a self-imposed restriction on the professionalism of buyers. Their appreciation of the commercial aspects would be improved if more account was taken of the specifications.

The case of misdirected interest shows the way in which some buyers give the impression that, although they are indeed interested in specifications, the nature of the content is too technical for them. This impression is conveyed by the way in which many will immediately pass all consideration of technical points over to engineers. This is often because, although they really are interested, they are much too busy with "commercial aspects"—a euphemism for the search for ever-lower prices.

It will often be found to be the case that when meetings take place between supplier and customer personnel to discuss and resolve specification problems, the buyer is conspicuous by his or her absence. This is unfortunate when one remembers how often one has been adjured not to enter into any commitment, or change in commitment, without the authority of the buying department. It is inevitable that the resolution of a technical problem will involve a change of commitment in one way or another. It must not be assumed, however, that there are not buyers who are knowledgeable in the manner suggested and are able to, and do, contribute greatly in engineering/technical discussions.

For too many buyers the subject of the preparation and use of specifications has been a closed book. This has meant that when specifications have to be sent out with an order, a check on the serial and issue numbers is the only check likely to be made. A new approach is needed to the entire subject of specifications, an approach that will extend the general professionalism of buyers and increase the effectiveness of the jobs. they do for their companies.

To begin with, one must consider the purpose of a specification, which is actually deceptively simple: "to communicate and to convey information" — information by means of which the recipient is informed of the wishes of the originator and enabled to conform to them. But buyers also need to know what the wishes of the originator are. They need to know because, after all, theirs is the responsibility for procuring that which the originator requires. To do this in a satisfactory, businesslike manner, it is necessary to understand the requirements—they must understand the specification(s). One must be clear about what one means by the word "understand" in this context. Websters' says "to know thoroughly, grasp or perceive clearly and fully...." *The Oxford English Dictionary* says "to have comprehension or understanding (in general or in a particular manner)." Meaning 1(b) in the *Oxford* seems to apply to the originator of a specification: "to be expert with, or at, by practice." The buyer should, therefore, have an understanding in a general manner.

Buyers are rarely involved in the work of preparing specifications. Yet this is an activity whose understanding could have a profound effect on the general complexity of a buyer's task and determine the degree of success which buyers may expect to achieve. Although there can be little doubt that many of the specifications

SPECIFICATIONS FOR BUYERS

that pass through buyers' hands are satisfactory, far too many are either badly prepared or badly written or both.

In one case a supplier company had about 5000 active customer specifications in its files, and it was known that at least 25% of them had given rise to problems. It is therefore highly desirable that buyers have an appreciation of the problems facing those who prepare specifications, as well as an understanding of the parallel problems facing the specification user. Knowledge of this kind will assist buyers in handling specifications and encourage them to question features they do not understand.

An extract from what one can reasonably claim to be one of the oldest specifications known will help to explain one type of problem. It is taken from the Old Testament of the Bible, Kings 1:6, describing Solomon's temple.

> In the inner sanctuary he made two cherubim of olive wood, each ten cubits high.
>
> Five cubits was the length of one wing of the cherub, and five cubits was the length of the other wing of the cherub.
>
> It was ten cubits from the tip on one wing to the tip of the other. The other cherub also measured ten cubits.
>
> Both cherubim had the same measurements and the same form.
>
> The height of one cherub was ten cubits and so was the height of the other cherub.

If a buyer had to buy a quantity of cherubim, he or she would unquestionably face a number of problems. As a specification it leaves quite a lot to the imagination. Perhaps the first point that becomes apparent is that no tolerances are given. This would greatly affect interchangeability on "bulk" orders. One should also know that, depending on the point in historical time and the geographical region concerned, the cubit varied between 18 and 22 inches.

So if more than one supplier was involved, as would often be the case today (not all of one's eggs in one basket), it is highly unlikely that two cherubim from different suppliers would be the same. The differences could be as much as a maximum of 3 feet in 15 feet—not very appropriate for a temple. (See the problem of the shirts in Chapter 3.)

The Old Testament contains another interesting example in Leviticus 1:3. It begins with a simple sentence that is very clear and specific:

> Whatever parts the hoof and is cloven footed and chews the cud, among the animals, ye may eat.

Unfortunately, however, the writer (or perhaps the translator) has blurred the clarity and precision of that statement by the following qualification:

> Nevertheless, amongst those, ye shall not eat these:
>
> The camel which parts the hoof and is cloven footed and chews the cud, is unclean to you.
>
> And the rock badger . . . is unclean to you;
>
> And the hare . . . is unclean to you;
>
> And the swine . . . is unclean to you.

I am not sure about all those exceptions, but the hare, at least, does not "part the hoof and is cloven footed." In this case clarity has been lost by a careless qualification—which is not an uncommon feature in modern specifications!

In the preparation of procurement specifications (sometimes called purchase specifications), and in any other specifications for that matter, careful consideration needs to be given to the content and to the manner of presentation. There are many rules that could be put forward, but there is one above all that should be followed:

SIMPLICITY

Neglect of this elementary rule probably causes more specification problems than any other.

A buyer who takes the trouble to read through a procurement specification of some size in detail may well be confused by the time he or she has finished reading it. There are seldom problems with short specifications, but that does not mean that short specifications never cause problems. However, it is likely that the buyer will form the opinion that whatever other considerations

SPECIFICATIONS FOR BUYERS

may have been in the mind of the writer, simplicity was not a primary concern. But the buyer will have a better idea of the nature of the problems that he or she could be passing on to suppliers. The buyer will also understand their reactions better and appreciate why so often they have to come back for guidance and clarification (If, of course, they choose to come back for guidance at all).

Consider the following extract taken from a specification issued to a supplier company by a large electronic equipment manufacturer and then try to imagine the thoughts that passed through the minds of those in the supplier company who had to try to understand what the customer had in mind.

> In order to obviate the possibility of subsequent modification and the high cost involved, the vendor is requested to make every effort to interpret the requirements of this specification in such a manner that the primary conception is correct.
>
> If a particular detail of the specification is not clear or is capable of more than one interpretation then the vendor must request clarification from. . . .

Quite apart from the fact that the user of a specification should *never* have to interpret a specification, let alone be invited to do so, the second sentence, completely contradicts the first in this respect. In any case, although the whole specification had to be read in detail to make this point clear, it was a fact that nowhere in it was there an explanation of that "primary concept" that the supplier was being urged to interpret correctly.

Can one then really be surprised if some of the strong feelings that are engendered by what can only be called "gibberish" sticks to buyers? After all, as far as the supplier is concerned, it is the buyer who has sent that specification. It only takes one example of this kind to do a great deal of damage to a carefully built up image.

Consider another example, one that certainly stretches the mind in terms of trying to understand what it means. It purports to be a definition of "rework."

> any processing or reprocessing operation, other than testing, applied to an individual component, or part thereof, and performed subsequent to the prescribed nonrepairing manufactur-

ing operations which are applicable to all components of that type at that stage.

Can there really be any more doubt that even if only for no other reason than self-interest, buyers must pay more attention to the specifications that pass through their hands on the way to suppliers? Properly used, specifications can be a powerful tool in the hands of buyers to help them to buy the products they need, at the right standard of quality and at the right price. But buyers must understand them and appreciate the nature of the information they are intended to convey.

When buyers read specifications, at least they attain a better understanding of the requirements concerning products they are buying. If they are confused by what they read, they may be sure that suppliers will also be confused. They should initiate appropriate internal action in their own companies to clear up all confusion before it reaches suppliers.

This brings us to a consideration of what I have called the "purchasing diamond." This is a concept that I developed some years ago for use in industrial training courses for buyers.

Advisory
Consultative
Familiarization
Foundation of authority
Extension of use
Knowledge
Action

The seven factors displayed in the purchasing diamond describe, in the proverbial nutshell, the extent of the interest in specifications that will be of benefit to buyers. It provides buyers with a powerful "right arm," even if that fact may not be immediately apparent. Essentially, the diamond is made up in the following way.

Advisory. Although in the case of specifications that are strictly for internal company use, it may not matter much, specifications for external use are another matter altogether. As a general rule, buyers have a very considerable store of knowledge about suppliers and about the what and wherefore of their products. It is in the best interests of specification writers, therefore, that whenever appropriate, they make use of this store of knowledge—in other words, seek the advice of the buyer!

Consultative. This factor is an extension of the first. Apart from the general advice that buyers can give specification writers, it is to the general advantage that they be regarded as a kind of internal consultant. That is, specification writers should look at buyers in this way and make use of the consultative aspect.

Familiarization. This facet is directed toward buyers and represents an action that they themselves take. They should make themselves at least reasonably familiar with the nature and content of the specifications that pass through their hands on the way to suppliers. It may be that some of the specifications are too voluminous, and perhaps too complex, for buyers to become completely familiar with them to the last detail. But they should at least be aware of the general nature of the contents and the principal requirements.

Foundation of authority. When a buyer places an order with a supplier, the buyer is also indicating to the supplier that he or she has knowledge of what it is that is being ordered. That is, the buyer is the authority about the order as far as the supplier is concerned. If problems arise as to the meaning of requirements in the specification and the supplier raises them with the buyer, in the first instance at least, a diminution in the authority of the buyer will result if he or she has to pass the supplier's inquiry on to someone else to resolve the problem. A very good example of this type of situation is described in Chapter 3. The consequences for the buyer concerned were somewhat unfortunate, but this was clearly his own fault. This case also demonstrates the need for application of the CAST concept, discussed in detail in Chapter 8.

Extension of use. Making use of these factors will result in an "extension of use" that the buyer has to his or her employer. The "boss" will know that the buyer is more than just a "buyer," and that can only help the buyer's prospects, as well as improving his or her general performance for the benefit of the company.

Knowledge. The store of knowledge that the buyer possesses will increase substantially, and this will help the buyer generally with his or her work.

Action. As a summation of the previous six factors, this final one indicates just what it is that the buyer can do. When it is necessary, the buyer can, and will, take action.

This, then, is the purchasing diamond, which helps to show that despite the fact that real diamonds may no longer be the investment wonder they once were, they are still a good friend to buyers.

3
What Is Being Ordered?

It is not at all unusual for a supplier to be somewhat confused about the specific nature of the item that they are supposed to be supplying on the basis of an order. Often, the order details are not as clear as they might be, and this includes association with related specifications. Some of the problems experienced by suppliers are discussed here together with some of the ways in which they may be overcome. Examples are given to provide positive guidance.

The first question that one must ask is : "What is it that is to be ordered?" It is a very simple question, but the answer is usually not quite as simple. It may be a button; it may be needles and thread; it may be men's shirts; it may be an airliner or a 10,000-ton ship. But whatever it may be, if you are an industry buyer buying in quantity, it will not be much good placing an order for, say, 10,000 buttons (one Hongkong specialist button manufacturer lists no fewer than 25,000 different buttons in his catalog). The order must be accompanied by a specification in sufficient detail to include, among other particulars, the physical dimensions, the color, and the number, size, and positions of the thread holes.

However, in industrial purchasing, orders of that simplicity are not very common, although I must confess a special liking for an order for "100 gross of 1 inch × No. 8 (screw gauge) csk. iron wood screws." That is about as simple as an industrial order can be. But the items that industrial buyers procure usually require more than that. They need a specification. One classic example of bulk buying will be described. The item was bought from a number of suppliers and the circumstances demonstrated the need for a good specification if the end customer, in this case the proverbial man in the street, was to be satisfied. The item? Shirts.

A number of years ago a large retail store chain bought shirts from several suppliers for sale under its own brand name. It soon became aware of a serious buildup in customer complaints about its shirts. For any particular neck size all the shirts were ostensibly the same, but an investigation revealed a most undesirable state of affairs. Not only were there variations in size between shirts of the same nominal size from different suppliers, but there were also significant differences between shirts of the same size from the same supplier. Not surprisingly, this was considered to be a most unsatisfactory state of affairs.

After a good deal of preliminary work, carried out jointly by the customer and the suppliers, the problem was solved by the preparation of a specification "book" of 20 or so pages for each type of shirt. All the requirements were set out: how, and between which points, measurements were to be made and the tolerances; the type of material to be used and how the various pieces of the shirt were to be cut; the needle and thread sizes; the number of stitches per inch; and so on. The end result of this work was a supply of shirts from the various suppliers which, within the allowed tolerances, were all the same size—a supply that, most important of all, gave customer satisfaction. This example was featured in a British government film about industrial quality.

The point about all this is that one must have an adequate specification for whatever it may be that is being purchased to ensure getting what is wanted. A specification that would be considered adequate can range from short and simple to long and complicated, depending on the nature and complexity of the item being ordered. A very good example of a specification that could be represented by a simple statement on an order is that of the wood screws mentioned earlier. On the other hand, it could be a book of many pages, rather like the shirts or even more so. I once received a set of four specifications for a single simple electronic component which had a total of 132 pages—most of which, it might be said, were quite redundant. It has been reported that the complete specification for the *Queen Elizabeth II* ocean liner contained more than 2000 pages, not counting drawings. An intermediate-sized example would be one for automotive pistons. In such a case the complete specification set might consist of a detailed drawing, a metallurgical specification, and an acceptance inspection specification.

Even when a specification is apparently satisfactory, it is highly desirable that the buyer make certain that the specification shipped is actually the right one. In one case a component supplier company received an enquiry from a regular customer for a large quantity of a number of different types of electronic components of commercial grade. The supplier was used to receiving orders from this customer for military grade only, so when the quotation was ready, a sales engineer took it in person to present to the customer's buyer. He also had a question for the buyer. He wondered if commercial-grade components were really what was wanted as it was normal for military grade to be supplied. The response from the buyer was more than a trifle frosty. If he had wanted military-grade components, he would have asked for them. However, the quotation was accepted and the order, a valuable one to the supplier, was delivered on time.

A few weeks later the managing director of the supplier company received a telephone call from the customer's chief engineer. He was in trouble; could the managing director come up and see him and discuss a possible way out of the difficulty? The problem was that the chief engineer had a large quantity of electronic components of commercial grade that were of no use to him, when what he needed was military grade. The error had resulted from a misreading of a purchase requisition by the buyer, who had not bothered to check with the engineers. A mutually acceptable resolution was achieved and the commercial components were returned and replaced by the equivalent military-grade components. By the way, the buyer in question did not remain with the customer very much longer.

Although it is important to ensure that specifications which leave nothing of consequence to the imagination of the supplier are available, it is of equal importance to ensure that the specification writers are not too clever. If this is the case, the result is likely to be a specification that may be ambiguous or confusing or both, and which will leave the supplier uncertain of the actual requirements. In another example of this kind an order was received by a supplier with a specification that covered two grades of the same component. The supplier was uncertain which grade was actually required, so supplied the one usually ordered by that customer. It was rejected as being "not in accordance with the requirements of the order." Obviously, it had to be the other grade. But surprise, surprise, that one was rejected for the same reason. The customer's own staff was also confused.

Once specifications of this kind are written and issued, it is not as easy to get errors corrected as one might think. In the particular case just described, the task of correction took no less than 12 months! It is better by far to be sure that specifications are above suspicion before they are issued. As problems of confusion are found, they should be cleared as quickly as possible, in the interests of delivery to schedule, good supplier relationships, achievement of the desired quality, and above all, containment of LOST COSTS. (The problem of lost costs is dealt with in more detail in Chapter 14.)

Cases of unclear specifications are found much more often than is generally appreciated and are usually found in large organizations with "systems." But even small companies are often involved in such problems. A further type of confusion arises when individual requirements are selected from existing specifications without consideration being given to their compatibility outside their original context and under new circumstances.

In retrospect many of these examples seem amusing. At the time, however, they were anything but funny, as every one cost the supplier lots of money. One should therefore do all that is possible to ensure that adequate specifications are available when required. If, and when, the buyer is completely satisfied that he or she knows what is to be ordered, one can be certain that the supplier will also be satisfied in this respect: the item ordered will be supplied as ordered.

Although this chapter has concentrated on the importance to buyers of knowing what is to be ordered, one must not forget the general tasks involved in the preparation of the specifications themselves. That is, the earlier and initial tasks of the design department. For readers who would like to delve a little more deeply into those activities and the responsibilities that go with them, Appendix 1 of this chapter presents a copy of a procedure that applies to the design department of Company A. These basic

WHAT IS BEING ORDERED?

responsibilities could be translated without too much difficulty to a totally different industry, say pharmaceuticals. It should be pointed out that the following paragraphs in the appendix are of direct relevance to the activities of the two purchasing departments of the company:
5.1.3.4, 5.1.3.4.2, 5.1.3.4.3, 5.1.5.3, 5.1.5.4, 5.1.8.4, 5.1.8.5, 5.1.8.6, 5.1.8.7, 5.1.9, 5.1.9.1, and 5.1.9.2.

APPENDIX 1: DESIGN RESPONSIBILITIES IN THE QUALITY SYSTEM

1. Purpose of the Procedure. To set out the responsibilities of the engineering department in the overall assurance of quality requirements in the products of the Company.

2. Scope. This procedure covers the activities of the engineering design manager and his or her department primarily, with lesser involvement on the part of the quality assurance, purchasing, and production departments. The lesser involvements are mainly of a liaison nature.

3. Related and Relevant Procedures and Other Documents

Def. Stan. 05-21, paragraphs 203, 204, 206, 207, 210
01-101-003   Control of Drawing Issues and Distribution

There are a number of other procedures that have not yet been written nor allocated numbers.

4. Responsibilities. The prime responsibility for ensuring conformance to the requirements of this procedure rests with the engineering design manager. However, the other departments mentioned in paragraph 2 do have responsibilities of a lesser nature to the extent of their involvement. In the particular case of the quality assurance department, where there is a liaison function, in some respects the responsibility is shared with the engineering department.

5. Requirements of the Procedure

   5.1. *Engineering Department*
      5.1.1. There are a number of specific areas of responsibility

and the following paragraphs set them out with guidance regarding methods of conformance in an approximate, at least, order of system involvement.

5.1.2. *Design Work Instructions*

5.1.2.1. Individual designers will have as many methods of approach to a design problem as there are designers. A wide range of approaches, especially in consideration of details, is likely to increase the probability of problems arising. The engineering design manager must, therefore, ensure that he or she has prepared for the guidance of the designers a series of detailed "work instructions."

5.1.2.2. These work instructions must depend to a considerable extent on the exact nature of the design work involved but should, as far as possible, require the designers to utilize standard items whose performance is adequately known and documented. For example, depending on the degree of criticality, there might be a set of departmental standards for fastenings. Another example might set out standard methods for laying out drawings to ensure consistent presentation of information; standard tolerances for particular purposes and methods of manufacture; and specific scales of size reduction in drawings for specific purposes.

5.1.2.3. *Codes of Practice*

5.1.2.3.1. A particular form of design work instruction is the code of practice. It may be convenient to use existing codes of practice, for example British Standard Codes, or specific codes can be produced to suit the particular requirements of the engineering department.

5.1.2.3.2. It must be recognized that codes of practice are not produced overnight, and compilation of data into a code will take place over long periods. Practically, a code prepared for internal use will never be fully completed, and this fact must be recognized.

5.1.3. *Design Reviews.* An important aspect of design work and one to which too little attention is given is that of design reviews. There are four phases of design reviews that must be considered.

5.1.3.1. *Phase 1 Reviews.* These concern the consideration that should be given to designs during the stages of their formulation and completion. Many aspects must be considered during these stages, of which the most important, perhaps, are ease of achievement of quality and reliability; suitability for manufacture with the manufacturing methods that are actually available; maintainability during the service life of the design; and the probability of achievement of the design parameters in actual service.

5.1.3.1.1. These reviews cannot be considered to be fully effective unless they are carried out in association with nominated liaison personnel from the other departments concerned. The engineering design manager must therefore ensure that those other departments nominate suitable personnel and that the reviews are made in association with those persons.

5.1.3.1.2. Being primarily responsible, the engineering design manager must ensure that proper records are prepared and maintained against particular designs in a suitable manner that will record the results of the review together with any actions that are to be carried out arising therefrom; and that follow up actions are taken which result in the outcome expected.

5.1.3.1.3. The engineering design manager must ensure that the design reviews are held at suitable intervals during the development of the design and in relation to the complexity of individual designs. At the very least the manager must ensure that there is at least a final review before the design is cleared for production.

5.1.3.1.4. A primary aim of the design reviews must be to ensure that when a design is cleared, the parties involved are mutually satisfied that the various design, quality and reliability, procurement, and manufacturing factors have all been dealt with in a satisfactory manner consistent with the economics of production and the demands of the design parameters.

5.1.3.2. *Phase 2 Reviews.* Phase 2 concerns the "continuous" review that should be given to designs after original clearance for production and during the duration of production and subsequent service in the field.

5.1.3.2.1. As in the case of phase 1 reviews, phase 2 reviews must be carried out in association with the nominated liaison personnel in the departments listed in paragraph 5.1.3.1, with the addition, in appropriate circumstances,of sales or marketing personnel for feedback from customers.

5.1.3.2.2. The engineering design manager must again determine the frequency of reviews, which should be based on a consideration of any problems that may arise during manufacturing or customer use. Depending on the nature of any problem that may arise, the manager may conduct a review with one or more of the nominated review personnel from the other departments.

5.1.3.2.3. The engineering design manager must ensure that similar records and actions are made and taken as are required for phase 1 reviews and stipulated in paragraphs 5.1.3.1.3, 5.1.3.1.4, and 5.1.3.1.5.

5.1.3.3. *Phase 3 Reviews.* These are carried out on new sets of drawings which are received from customers on the initiation of a new contract. The engineering design manager must ensure that they are carefully reviewed in the department, and in association with the liaison personnel from the other departments, including QA, as necessary. These reviews serve a number of purposes, of which the following are examples.

5.1.3.3.1. To determine that all necessary information to carry out the contract has been received.

5.1.3.3.2. To determine that the information that has been supplied is consistent with the requirements and that there are no anomalous situations.

5.1.3.3.3. To determine if any new manufacturing techniques or new gauging methods might be required, which would then have to be developed.

5.1.3.4. *Phase 4 Reviews.* As the engineering design manager is responsible for any designs that may be prepared by vendors, it is of particular importance that he or she institutes suitable procedures to ensure that there is a continuous arrangement for the review of such designs.

5.1.3.4.1. It is of importance that the engineering design manager ensures that the buying department be kept fully informed of any arrangements he or she considers necessary in such cases. The manager must also ensure that the QA liaison engineer is kept informed of the

WHAT IS BEING ORDERED? 31

progress of such design work, just as is the case with in-house design work.

5.1.3.4.2. The engineering design manager must ensure that suitable records are maintained of the reviews of vendor designs, and any actions resulting from them, in a manner similar to that for records maintained for phase 1, 2, and 3 in-house reviews.

5.1.3.4.3. The engineering design manager must ensure that vendor design departments use suitable work instructions, codes of practice, and liaison techniques during the course of any design work undertaken for the Company.

5.1.4. *Design/QA Liaison*

5.1.4.1. In consultation with the quality assurance manager, the engineering design manager must ensure that the quality department appoint a liaison engineer who will work with the designers in the reviews described in paragraphs 5.1.3.1 and 5.1.3.2 and their subparagraphs.

5.1.4.2. In certain circumstances it may be necessary to appoint more than one QA liaison engineer, with each having QA liaison responsibilities for specific projects.

5.1.4.3. The designer should expect that the QA liaison engineer will report on any problem in the quality and reliability field that is found to arise during the course of manufacturing or testing or during actual customer service.

5.1.4.4. The designer should also expect that the QA liaison engineer, with his or her acquired knowledge of the projects on which the designer is working, will bring to the designer's attention problems arising on production designs which he or she considers might also arise on subsequent designs unless suitable design corrective action is taken to prevent its happening.

5.1.4.5. Liaison is a two-way system of cooperation, and the QA liaison engineer must, in turn, expect that the designer will seek his or her opinion and be prepared to use his or her knowledge about problems that arise during the course of the designer's work on a project.

5.1.5. *Design Corrective Actions*

5.1.5.1. The engineering design manager must ensure that suitable documented action is taken to correct any deviant suitable documented action is taken to correct any deviant situation that is found to exist during any review activity.

5.1.5.2. Corrective actions that must be undertaken under such circumstances may require modification action on drawings/specifications for products at the manufacturing stage. In those instances the engineering design manager must ensure that the modification committee (whose work is described in another operating procedure) is informed at the earliest opportunity.

5.1.5.3. In particular, the engineering design manager must ensure that in any instance in which a vendor design organization undertakes design work on the manager's behalf, constraints are placed on that vendor design organization similar to those that operate in the manager's department. The manager must also ensure that there is an adequate flow of information to him or her from the vendor and that this information is made available to the QA liaison engineer(s). The manager must also ensure that designs and design changes that are initiated by a vendor are approved by him or her and that they are suitably recorded.

5.1.5.4. Adequate records are to be maintained and the engineering design manager must ensure that the records are readily available to the designers, and that the designers are made aware of the fact that it is their responsibility to make use of these records and the information contained in them.

5.1.6. *Modification Control*

5.1.6.1. The operation of a system for the control of modifications is dealt with in detail in another operating procedure, but it is the responsibility of the engineering design manager to ensure that the designers carry out these requirements in all cases in which it is necessary to make modifications or changes to drawings or specifications. The responsibility extends to ensuring that all necessary consultations are made with any relevant department or individual or customer or other body.

5.1.7. *Control of the Preparation, Issue, and Distribution of Drawings*

5.1.7.1. In this context, "drawing" must be taken to include any drawing, specification, technical instruction, or data of

WHAT IS BEING ORDERED? 33

a similar nature that may be issued by the engineering department to facilitate the manufacture of a product.

5.1.7.2. There is another operating procedure that covers this requirement in detail, but it is the responsibility of the engineering design manager to ensure that insofar as his or her department is concerned, these requirements are followed.

5.1.8. *Design Responsibility for Proprietary Product*

5.1.8.1. In those designs in which the designer makes use of proprietary products, it is a responsibility of the engineering design manager to ensure that the designers properly evaluate proprietary products which they propose to use, to ensure that the products are fully capable of meeting the standard of performance expected of them.

5.1.8.2. The complexity of design of proprietary products used may range from simple items that are easy to supply in quality meeting a national standard, to complex items of electronic equipment required for control systems.

5.1.8.3. The method of evaluation will vary according to the simplicity or complexity of the product being considered, but the engineering design manager must ensure that it is suitable for the product concerned. The manager must also ensure that the details of the evaluation are suitably recorded so that it will be possible to review the original work later, when the actual performance of the product in service can be compared with the result of the original evaluation.

5.1.8.4. Records of evaluations that have been made should be available for future reference when consideration is being given to the inclusion of similar proprietary products in other new designs.

5.1.8.5. Once evaluations have been made of suitable proprietary products and decisions made as to their suitability, it is a further responsibility of the engineering design manager to ensure that the managers of the two purchasing departments are informed of the types and manufacturers of the proprietary products that have been evaluated and approved by the engineering department so that only approved items may be purchased.

5.1.8.6. The engineering design manager must also ensure that when design work is being done for them by a vendor, the vendor suitably evaluates proprietary products in accordance with the manager's requirements, and makes suitable records. Both of these requirements are as in paragraph 5.1.8.2. The manager must also require that the vendor inform him or her of any evaluation work done on proprietary products and of the results of such evaluations, and that the vendor's purchasing department has been informed as in paragraph 5.1.8.5.

5.1.8.7. The engineering design manager must either carry out a subsequent review of the vendor-provided proprietary products, or ensure that the vendor carries out such a review, as outlined in paragraph 5.1.8.3. Should this review indicate that the products are not performing as expected, the engineering design manager must take suitable corrective action.

5.1.9. *Design Data for Procurement*

5.1.9.1. Although the actual provision of drawings to the purchasing departments for procurement purposes is the direct responsibility of the design division libraries under the control of the chief planners, the engineering design manager must ensure that designs produced for those products which are to be purchased take this fact into account and that the data contained in them are suitable for that purpose. That is, drawings for product that is going to be bought outside may need to contain more information than would be the case if it were to be manufactured in-house.

5.1.9.2. In certain instances, decisions will be made that product which would have been made in-house, and product which has been partially machined in-house, will be made out or finished out. The liaison of the engineering design manager with the chief planners should be extended to include the production control departments to ensure that the manager is made aware of those instances as well. When they arise, the manager must review the drawings concerned to determine if additional information should be added to them to assist the vendors.

5.1.10. *Other Factors*

5.1.10.1. The engineering design manager must ensure that the design staff members are aware of their individual responsibilities for product quality and reliability, which, basically, are determined at the design stage. Account must be taken of economic factors and manufacturing capabilities, among other things, in determining dimensional tolerances.

5.1.10.2. Any other factors that might contribute to the economic effectiveness of the designs should also be considered, and the designers must be encouraged to take them into consideration (e.g., value engineering).

5.1.10.3. The engineering design manager should ensure that when consideration is being given to the causes of nonconformance, opportunities are made available to the designers to participate in such consideration: both to assist in the consideration and to take advantage of any relevant information that may affect their design work.

4
The Order

Generally speaking, the preparation of orders is a routine task for buyers. But when one looks at an order from the point of view of quality, one finds that a rather different outlook is necessary. First it is accepted that the average supplier, regardless of the size of the company concerned, is anxious to give customers just what the customers want, no more and no less. If there is anything about the order that in any way tends to confuse the supplier, it is going to cost money to clarify and will reduce the likelihood that the customer will receive what is wanted. The buyer has to make sure that the supplier is given clear instructions which specify precisely that which is wanted. In some, perhaps many cases, all that is necessary is a simple statement on the order, rather like the example of the woodscrews given in Chapter 1. But in many other cases there will be a need for more information to be given on the order. This brings up the corollary to that premise: There must not be *too* much information.

Consider the following example:

130 Bridge Rectifiers 1B40K40
Part No. 77015438
Drawing No. 817494, Sht. 27, Iss. 77
Drawing No. 819624, Sht. 5, Iss. 1

All nice and straightforward with clear and precise information for the supplier? No. The 1B40K40 was perfectly satisfactory as it was the supplier's own reference number. But the supplier did not have any information about the three remaining references: that is the part number and the two drawing numbers. As a good supplier, anxious to give the customer just what is wanted, the supplier's sales department telephoned the customer's buyer to say that they did not have any information about the last three references on the order. The response from the buyer was somewhat surprising: "Don't worry. You don't need them." This both puzzled and worried the supplier's quality manager. Why would the customer put totally irrelevant information on the order? He felt that he needed to know more about the reason this information was considered unnecessary, although on the order. He thus telephoned the customer's buyer to ask him just why the information was not of use. The answer was a little surprising and the explanation was as follows:

> Part No. 77015438. This was the reference number for that particular item in the customer's reference catalog of purchased items. *Clearly not required by the supplier.*
> Drawing No. 817494, Sht. 27, Iss. 77. The customer's mechanical drawing office prepared drawings with the dimensional information on all parts. This drawing is a compendium set covering this type of component, and sheet 27 was the one for this particular component. The high issue number, quite unheard of previously, merely indicated the 77th change to the entire compendium set. *Clearly not required by the supplier.*
> Drawing No. 819624, Sht. 5, Iss. 1. Except that this contained the electrical details and was prepared by the electrical drawing office, this is the same kind of information as that provided by the mechanical drawing office. *Clearly not required by the supplier.*

The point about all this is that ambiguous and/or confusing information on an order will, directly or indirectly, have some effect on the quality of the service that will be provided by the supplier. Of more direct influence on the quality offered by the supplier can be the details that the order conditions should include relating to the level of quality of product or service required. It is often the case that order conditions do not include any stipulation about quality.

THE ORDER

Long experience has shown that very often there is little or no guidance in the conditions of the order which relate to the quality of the material or service being ordered. A major exception, of course, happens in the case of military orders.

A number of years ago, to deal with a similar situation in which there were no order conditions relating to quality, and where embarrassing problems had arisen, I prepared two sets of "quality clauses" for inclusion in order conditions. The first was intended for routine, use, as may be judged from its title "Model Inspection Conditions Clause" and from the contents.

1. It is a condition of supply against this order that acceptance, or rejection, of any shipment may depend on the result of a sample inspection, using a standard sampling inspection plan to stated acceptable quality levels and to an acceptance specification.
2. The acceptable quality levels and the acceptance specification shall have been agreed in advance between the buyer and the seller and the specification will identify those features that should be inspected and indicate their relative importance.
3. The buyer reserves the right to waive any or all of the inspection requirements and to alter the acceptance specification from time to time. However, an altered specification may be used for rejection purposes only after mutual agreement between the buyer and the seller.
4. The buyer reserves the right to use any shipment rejected at sampling inspection. The supplier may also request reconsideration of any rejection.
5. The buyer reserves the right to negotiate a revised lower price for any substandard supplies when a valid rejection decision is reversed.

The second of these two "quality clauses" was called a "special condition of order."

1. Shipments against this order must each be covered by a certificate of confirmed quality signed by your quality manager.
2. The Certificate must state that the shipment fully conforms to the specification agreed between the buyer and the seller and was accepted accordingly.

3. The buyer reserves the right to inspect any shipment to the agreed specification if desired and to pass or reject the shipment depending on the results of such examination.

The second clause represents the achievement, or the attempt to achieve, one aim that should be in the forefront of every buyer's mind: reduction of the cost of procurement by the elimination of the need to carry out any inspection of incoming supplies. It must also be remembered that this is an aim that can be achieved only through the close cooperation of the buyer and the buyer's colleagues, and by making use of other quality techniques discussed in later chapters.

It must also be pointed out that this aim will not be achieved easily or without a great deal of perseverance and effort on the part of the buyer and many of his colleagues. In my own experience, in one case over a period of two and a half years, only two suppliers out of more than 200 were brought up to this standard. How one may set about the task of practical achievement of this aim is dealt with fully in later chapters. However, it is hoped that enough has been said to make it clear that when one considers the required quality aspect of procurement, there is more to an order than may at first sight seem to be the case!

Before leaving the subject of the order, two more points should be mentioned. For the first, consideration should be given to other information that may have to be conveyed to a supplier by means of the order. That may include special requirements with regard to documentation or packaging or method of shipping. For the second, one must be certain that control is exercised over the possibilities that exist in a variety of ways for changes in the requirements of an order. It is so easy for representatives of a customer company to agree, or request, changes in an order which the supplier will accept as binding but which are not covered by the necessary order amendment. The consequences of breaches in standard procedures can be considerable, and buyers are well advised to be on their guard.

Although the foregoing discussion is more likely to be of direct interest to buyers in smaller companies, buyers in larger companies should not think that it is not for them. All the examples quoted came from large companies. However, there is no doubt that in more recent times buyers in the larger companies and

THE ORDER

purchasing organizations have been more likely to follow more formalized procedures, such as those required by British Standard BS 5750 or related or successor standards (see Appendix A). But it should again be pointed out that adherence to BS 5750 or its equivalents will not prevent errors from arising.

Following are two paragraphs from the January 1986 issue of the newsletter *Random Sample*, published by the Santa Clara Valley section of the American Society for Quality Control.

> Those of us who have worked with vendors have sometimes wondered how certain suppliers make consistently good quality products at competitive prices, but their quality assurance systems are documented in a sketchy manner and lack depth and thoroughness. These are usually small sized companies, and may have changed management hands recently. On the other hand, we have also worked with medium to large sized companies where they have elaborate and impressive quality assurance systems, but the quality of the final product lacks consistency and is subject to occasional breakdown.
>
> The inescapable conclusion is that an excellent quality system plan does not ensure a final quality product or quality service. If the departments and individuals in the organization find ways and excuses for not following the prescribed plan, then the overall quality assurance plan only serves to provide a false sense of security, and the consequences could, in a short time, be fatal to the survival of the entity.

How true!

5
Vendor Selection and Approval

To begin with, buyers who are in the market for supplies of one kind or another are likely to find that there will be a number of possible vendors for what they want to buy. At first glance it is also likely that some, if not all, of them will appear to be reasonable possibles, so that buyers will have a decision to make. With which vendor should they place their orders? The commercial considerations must be equally satisfactory, of course, but there is another factor that some buyers will have to take into account: the reluctance of some vendors, usually large ones, to accept orders for less than a stated value, which could be quite large. Associated with this minimum-order value factor is a parallel factor: If buyers are likely to be buying small quantities of one or more items and/or with frequent or small deliveries, consideration should be given to the use of stockists or distributors for those items if at all possible.

There was a time not so long ago when vendor selection might depend on the fact that a buyer played golf or had lunch with the vendor's manager or salesperson. Today there are more practical methods for selecting a suitable vendor.

There are several ways in which one may select and approve vendors, and they range from the simple to the relatively complex. In this chapter two methods are discussed. The first is quite simple and has been known to me for many years. It is now used by, and likely to be of particular value to, a smaller company. The second method is one of worldwide interest, based on a series of national standards for quality systems. It is used almost exclusively by larger companies and purchasing organizations, especially governmental agencies. But first, the simple method.

It is usually the buyer who decides that he or she has an interest in considering a possible new vendor. There will be preparatory work to be undertaken when that decision has been made. Apart from the normal buying tasks there will be much of an investigative nature, later, for the buyer's colleagues. The overall outcome will provide the necessary reassurance to the buyer that a new "prospect" will deliver the quality level desired and will do so in a consistent manner. The quality investigative work is not begun, however, until the buyer has formally requested the cooperation of quality colleagues in an assessment exercise. The tasks that the buyer will have carried out include the usual commercial points: that the product is available; that there is capacity to meet volume needs; that the vendor's lead times are acceptable; and that suggested prices are acceptable for further negotiation and not likely to change once negotiated.

It is very helpful when an exercise of this kind is under way if an individual buyer and a quality surveyor can be assigned as a team to look after a specific vendor project. (See Chapter 12 for details on the role of the quality surveyor.) Apart from helping to improve the buying/quality relationships, this will encourage the extra-close cooperation that is so essential when assessment exercises of any kind are being carried out.

The basis of the simple system is a questionnaire that is prepared jointly by the buyer and the quality surveyor of the customer company. It may be of interest to note that a comparison between a vendor questionnaire prepared almost a quarter of a century ago and a much more recent one showed few differences. That suggests that the older one, prepared when the "science" of vendor selection was in its infancy, must have been soundly prepared. A typical set of titles for the various sections of such a questionnaire follows.

1. General company information
2. Quality organization, responsibilities, and details
3. Purchasing system
4. Available metallurgical and chemical laboratory facilities

VENDOR SELECTION AND APPROVAL

5. Available nondestructive test facilities
6. List of the various sections of the quality department
7. Detail of incoming material controls
8. Detail of the process control systems in use
9. Methods for handling, segregation, and identification of materials
10. Detail of audit/inspection functions and stages at which used
11. Detail and extent of controls over inspection equipment
12. Detail of standards laboratory, or equivalent facility
13. Detail of quality manual, if any
14. Effectiveness of housekeeping

Although prepared jointly, this questionnaire should be completed with care. Due regard should be paid to the nature of the proposed vendor and the size of both companies, and whenever it is thought to be necessary, changes should be made from any previous version. The information it is intended to elicit will form the basis for a detailed examination of the vendor's capabilities, whether or not that takes the form of a visit. The responses to such a questionnaire are sometimes themselves sufficient for a decision to be taken.

Most of the questionnaire will be prepared by the quality surveyor, part of it will be a joint responsibility, and part will be primarily the responsibility of the buyer. Following is a suggested set of questions that might form some, at least, of section 3 just proposed, which deals with the purchasing system.

1. Do you have a formal procedure for selecting your suppliers?
2. Do you require your suppliers to operate suitable systems for the control of their product quality?
3. Do you examine, in any way, the quality systems of your suppliers?
4. Do your purchase orders contain all the technical information that would be required by your suppliers so that they will be able to supply to your requirements?
5. Does your purchasing department liaise with your quality department to ensure that your quality requirements are stated correctly and fully on your purchase orders and other purchasing documents?

When it is ready, the questionnaire should be sent to the prospective vendor under cover of a letter from the buyer, explaining the general purpose of the exercise. The letter should also say that depending on the nature of the answers, which will be carefully considered, it may be necessary for a visit to be paid to the vendor by the buyer and a quality surveyor. To be sure that proper attention has been paid to the questionnaire, the buyer should also request that either the chief executive or manager of the vendor sign it upon completion.

Upon its return, the questionnaire will be carefully examined by the buying/quality team. From this examination the scope and depth of the assessment to be made will be determined. In the case of simple products it may be decided that the answers given in the questionnaire contain sufficient information, of the right kind, for an immediate decision to be made about the suitability of the vendor for the item(s) that the buyer wants to order.

In the alternative case, decisions will be made as to the broad strategy for the assessment visit and on the composition of the team that will make the visit and its duration.(See any of the Quality Assurance System Standards listed in Appendix A.) Obviously, one of the major considerations in making these decisions will be the relative complexity of the item(s) it is intended will be ordered and the size of the vendor company. The answers will vary from a single member—usually a quality surveyor—for a half-day, to a small team of two for one or two days.

When a decision to visit has been made, the buyer should make the arrangements with the vendor. The buyer will advise the vendor of the number of people making the visit and the expected duration. This information is very important to the vendor. It will enable him to determine just which of his staff members will need to be made available to act as guides for the visitors and what accommodations have to be provided for private team consultations. The buyer will have requested that accommodations and other necessary "domestic" arrangements be made; that is meals, and refreshments.

At this point it is important to mention that the assessment visitors should form as good a relationship as possible with those members of the staff of the vendor with whom they come into close contact. Their task will be made much easier as a result. These contacts, incidentally, will not be confined to members of

VENDOR SELECTION AND APPROVAL 47

the quality department of the vendor. While the buyer is discussing the commercial aspects of the proposed order(s) with the commercial and sales staff of the vendor, he or she will also be forming opinions of the vendor on a commercial basis. The buyer's quality colleague, will be devoting his or her efforts to the technical/quality aspects. The next few paragraphs will discuss these in some detail as they are of considerable importance to the whole exercise.

The task of a quality surveyor working with a questionnaire such as the one discussed earlier is relatively simple and, ordinarily, the person should have little difficulty in completing the task in one day or even less. During this time the surveyor should explain to the vendor's quality personnel his or her company's basic quality requirements and the reasons for them. If the company has prepared a vendor quality manual, a copy should have been sent to the vendor in advance of the visit. In this way the vendor's quality personnel will have had an opportunity to familiarize themselves with the requirements in advance and prepare any questions they may have about aspects of the customer's quality policy about which they feel unsure or need more clarification.

Following is a list of the titles of the principal sections in an "incoming controls manual" published by an American multinational corporation which is a word leader in its field. The complete manual runs to 42 pages.

1. Incoming Controls: General
2. Destructive Tests
3. Forms and Records (this is the largest section)
4. Terminology
5. Care and Use of Gauges and Equipment
6. Tolerances
7. Appendix

If the vendor is going to be carrying out work involving metal finishing and/or treatment processes, arrangements should have been made for a laboratory staff member to accompany the team. These processes are very specialized and it's possible that the quality surveyor might not be sufficiently knowledgeable about them.

Apart from questions the vendor may have prepared in advance of the visit, the quality surveyor should be prepared, if he or she thinks it necessary, to elaborate on aspects of the company's quality requirements. This may, particularly, be the case with regard to sampling inspection and its implications for the vendor. This ought not to be the case, but a surprisingly high proportion of industrial companies do not fully understand, if in fact they understand at all, just what happens when sampling inspection decisions are taken on the basis of only a sample of a shipment. Despite all the information which there has been publicly available for more than half a century, the amount of ignorance on this subject is alarming.

In 1957, I carried out a survey of a group of 200 vendors to the British plant of a well-known American company, to determine the extent of their understanding of the nature and implications of sampling inspection. Perhaps all that need be said about the outcome of the survey is that a special booklet was prepared that was distributed to all 200 vendors, called "A Simple Guide to Sampling Inspection." In the late 1970s a similar survey was carried out and the results were very nearly the same. Despite all the public information, there seemed to have been little advance in general knowledge on the subject of sampling inspection on the part of industry in general over two decades. As its guidance still seems to be necessary, a copy of this guide is reproduced in Appendix 1 of this chapter.

For this reason, the quality surveyor should be prepared to go into considerable detail about his or her company's sampling inspections procedures and about the implications that there can be for a vendor in the event of the nonacceptance/rejection of a sample. He or she may also have to provide details of sampling plans and sources from which details can be obtained. It would not come amiss to have available copies of BS 6001,"Sampling Procedures for Inspection by Attributes" or MIL-STD-105D , the American military equivalent. These are by far the best known sampling plan standards throughout the world.

The scope and methods of control considered to be satisfactory for one class of product may be quite different from that considered necessary for another class of product. In such a case, if the scope and methods are considered to be acceptable for the item under consideration, the fact that it may not be satisfactory for other products being made should be ignored.

During the course of the visit the buyer and the quality surveyor

VENDOR SELECTION AND APPROVAL 49

will have made notes of their observations and the comments by the vendor's staff and all generally related to the questionnaire. After return to their own company the final report will be prepared, usually by the quality surveyor in collaboration with the buyer. However, it should have been possible to give the vendor at least an interim decision at a final meeting before leaving his factory. Upon completion of the report it will be considered carefully by the buying/quality team and, either the interim decision will be confirmed or in exceptional circumstances, a different decision will be reached. There are, in fact, four possible outcomes.

1. There is complete satisfaction and formal confirmation of approval.
2. There is general satisfaction except for one or two minor points. These will not prevent formal approval from being given, although the supplier will be requested to take note of them and take suitable corrective action.
3. The results do not justify immediate approval being given, for reasons that are stated. However, if the supplier will initiate suitable corrective action within a reasonable time, a further visit will be made on a mutually agreed upon date when it would be expected that an approval could then be given.
4. The results are very unsatisfactory for reason that are stated. The vendor will be informed that there seems little prospect of approval unless there were to be dramatic improvements.

In the remainder of this chapter are discussed what might be called "modern" systems of vendor assessment. They are almost all based on different national standards, depending on the country in which one is working. In the United Kingdom it is BS 5750, in Australia, Australian Standards 1821 to 1823, in Canada, Z299, and now a world standard, ISO 9000 and its subsidiary standards, which, it is expected, will replace all the individual national standards (see Appendix A). However, these are all civil standards. In the military field, in which most of the work on assessment was originally carried out, there are separate standards. In Britain there is the 05-21 to 05-29 series of Defence Standards (Def. Stans.) and in the United States there is the long established MIL-Q-9858A (1963) (see Appendix A). I have been shown a proposed revision of 9858, dated 1976, which from a quick scan looked very similar to the UK Ministry of Defence (MOD) series of Def. Stans. 05-21, -24, and -29. However, apparently, this proposed revision was not proceeded with. These MOD standards are all based on NATO (North Atlantic Treaty Organization) Allied Quality Assurance Procedures (AQAPs) (see Appendix A).

A decision was taken in 1985 by the U.K. MOD to use the AQAPs instead of either their own Def. Stans. or BS 5750. A prime reason was the fact that they are internationally recognized and used under NATO standardization agreements (STANAGS) as well as by other governments in several countries so that for defence contractors working to their requirements there will be international recognition of approval. The AQAPs were invoked progressively in MOD contracts from September 2, 1985. From that date contractors registered under Def. Stan. 05-21 will be deemed to satisfy the requirements of AQAP-1 Edition 3 until the next due assessment. Contractor assessment against AQAP-1 Edition 3 was, after assessor training, introduced on January 2, 1986.

AQAPs 4 to 9 are the same as the equivalent Def. Stans. 05-24 to 05-29. Except for 05-28, which is an Anglicized version of AQAP 8, the rest of the 05-20 series of Def. Stans. will be withdrawn in due course. For the guidance of readers the text of AQAP-4 Edition 2 is reproduced in Appendix 4 of this chapter. This deals with the "middle" level of requirements for which, probably, at least 70% of companies seek approval. AQAP-5, incidentally, not only contains the requirements of AQAP-4, it is also a complete guide to its implementation.

As a generalization it could be said that all of the basic requirement standards, except for MIL-Q-9858A, are broadly similar in the following way. They all describe a series of basic quality systems which, it is considered, will cover all needs. Usually, there are three of these systems, except in the case of Canadian national standard Z299, with four. For general convenience I will continue the rest of this discussion on the basis of the requirements of BS 5750, although there is little real difference between it, the ISO 9000 series, the British Def. Stans. or the AQAPs, either, for that matter; (see Appendix A). Indeed, there is little real difference between any of these standards when one gets down to basics.

It is highly desirable that buyers possess at least a nodding acquaintance with the contents of whichever is the appropriate

VENDOR SELECTION AND APPROVAL 51

assessment standard for quality system approval in their country. It is likely that buyers in the larger companies and organizations will know about them and that those companies and organizations will have organized themselves to handle whichever quality system level they consider appropriate for their company. Those in smaller companies may not know much about them. But most industrial companies are small ones and their buyers may not possess enough knowledge to judge the situation. The following paragraphs provide some detail about the three levels of systems that are generally recognized. These are levels 1, 2, and 3, which are dealt with in Parts 1, 2, and 3, respectively, of BS 5750. the standard also has Parts 4, 5, and 6, which are guidance texts for Parts 1, 2, and 3, respectively. The basic requirements in the first three parts are outlined next.

British Standard 5750

Part 1. This is the highest level and applies when there is specification of design, manufacture, and installation. It specifies the quality system to be applied when the technical requirements of materiel and/or services are specified principally in terms of the performance required or, where design has not yet been established. In these circumstance the supplier is frequently responsible for design, development and manufacture, field trials and installation work. Reliability and other characteristics can be ensured only by control of quality throughout all phases of the work.

Part 2. This is what might be called the median level, which will probably apply to about 70% of manufacturing companies. It applies when there is specification for manufacture and installation, only. It specifies the quality system to be applied when the technical requirements of the materiel and/or services are specified in terms of established design and where conformance to specified requirements can be ensured by inspection and test during manufacture and, as appropriate, installation.

Part 3. This is the lowest level and applies when there is specification for final inspection and test only. It specifies the quality system to be applied when conformance to specified requirements of the materiel and/or services can adequately be established by inspection and tests conducted on the finished materiel and services.

The foregoing descriptions have been taken from BS 5750 itself.

Parts 4, 5, and 6. These three parts are intended to provide a good understanding of the standard itself as well as providing assistance in applying it. The guidance is not exhaustive, but it does highlight all the important aspects to which attention should be given when one is examining the operating quality system of a manufacturing company. This it does mainly by going through the requirements in sequence and providing examples of what one should expect. There is one aspect of the guidance which is, perhaps, not quite as clear as one might have wished. Generally, all the examples that are discussed in the guidance notes relate to large organizations. Thus when one is applying the standard to smaller organizations, some care has to be exercised to ensure that one is not too rigid in applying the requirements. See also the self-determination program, presented in Appendix 5 of this chapter.

A broad list of the requirements from the middle level of BS 5750 is given below and will provide a better idea of what is expected.

1. Quality system
2. Organization
3. Review of the quality system
4. Work instructions
5. Records
6. Corrective action
7. Documentation and change control
8. Control of inspection, measuring and test equipment
9. Control of purchased materiel and services
10. Manufacturing control
11. Purchaser supplied materiel
12. Completed item inspection and test
13. Sampling procedures
14. Control of non-conforming materiel
15. Indication of inspection status
16. Protection and preservation of product quality
17. Training

VENDOR SELECTION AND APPROVAL 53

(Readers will see from a study of Appendix 4 that there is not much real difference between BS 5750 Part 2 and AQAP-4.)

When one is considering using a vendor who will be assessed to 5750, or any of the equivalents in Appendix A, it is desirable that they possess a quality manual, at least in the case of levels 1 and 2. Whether or not a manual exists in those words is in a sense unimportant, as the actual requirement is that the "contractor" have detailed written procedures. It could be said that a manual is just a "collected edition" of written procedures. However, those procedures should cover all the significant functions required for the overall control of quality and are as called up by the relevant part of BS 5750. In a rather simple sense, assessment by this method could be regarded as a fairly extensive development of the simple questionnaire method described earlier in the chapter.

When assessments are carried out in accordance with these standards, there would be little doubt but that the task would be carried out by a team of assessors (as they are now called), who could number up to four or five members, under a team leader. The visit is likely to last for a minimum of two days and could take up to a week, although this obviously depends on the size of the vendor being assessed and the general complexity of the product under consideration. It would be planned in detail in advance by the team leader, who may have paid a preliminary visit to the proposed vendor during which he or she would discuss with the vendor's chief executive the arrangements that the team leader would like to have made. An important task during such an advance visit would be to obtain a copy of the quality manual/written procedures on which most of the advance planning and allocation of tasks to individual assessors would be based.

During the assessment, the actual operation of a number of the procedures will be examined in detail in the presence of a member of the vendor's own staff. Any deviations found are immediately noted and must be agreed to by the vendor guide. At the end of the assessment there will be a meeting between the team and those of the vendor's staff involved, including, of course, the chief executive. During this meeting the team leader will discuss the results that have been noted and the conclusions reached. If minor deviations have been found during the visit and there has been an opportunity to correct them, that is usually considered to be acceptable. If other than minor deviations have been found, suitable remedial actions will also be agreed to with a time scale for their implementation together with a date for a further, small-scale visit to confirm the implementation. Before the end of the final meeting, however, it is essential that if at all possible, the assessment decision is given to the vendor. ANSI/ASQC Q1/1986 also deals with the subject of vendor assessment (see Appendix 2 of this chapter).

When a subsequent visit has to be made, it would be expected that normally, the vendor would receive approval. Unfortunately, this does not always happen. In one case I know about, a particular vendor failed to gain approval after three successive visits. It seemed clear to me that the prime cause of the failures was a lack of management appreciation of the need for serious planning and commitment when preparing for an assessment.

One important difference between the two methods of assessment of vendors which have been described is that in the second there is no grading of vendors other than what might be conveyed by an approval at one of the three levels available. A vendor is either approved or is not approved. If approved, the vendor is deemed to be fully capable of supplying the range of product in complete conformance with the requirements of the relevant specification.

In this chapter a good deal of space has been devoted to a description of the "simple" method of assessment of vendors for a number of reasons:

1. There is much more information available about "modern" methods than for the "simple" method.
2. By far the largest number of industrial companies are small companies, and for the small company, the "simple" method may be the most convenient, and it is easy to apply.
3. It can be even more convenient for large companies to use it.
4. It has resulted in friendlier relationships between customer and vendor companies.

There are five appendixes to this chapter and they contain the texts of the following documents:

1. A booklet prepared nearly 30 years ago to provide vendors with guidance about sampling inspection: the "why," "how,"

VENDOR SELECTION AND APPROVAL

and "what" of what happens to their product when it is delivered to the customer. Regrettably, it still appears to be necessary, although some minor changes to the text are required.
2. For Company A, the procedure for assessment, selection, and control of vendors.
3. For Company A, the procedure detailing requirements for vendor gauge control and calibration.
4. NATO Inspection System Requirements for Industry, AQAP-4, Ed. 2.
5. A self-determination program for compliance to BS 5750 or equivalent national or international standard.

APPENDIX 1: A SIMPLE GUIDE TO SAMPLING INSPECTION *

Purchase Parts Inspection (PPI)

Purchase parts inspection (receiving or goods inwards inspection) is responsible for maintaining the engineering and quality control standards on all finished parts purchased by the Company. With this in view, PPI must constantly strive to maintain the desired high quality at the lowest operating cost consistent with good inspection. This aim is to be achieved in two ways. First by ensuring that, through the medium of the Company's purchasing department—this is essential—close liaison is maintained with all vendors. This liaison must ensure that all vendors are made aware of our quality standards, our inspection procedures, and our gauging methods. Second, vendors need to be told that requests for information about these matters are welcome and that the quality control division is anxious to have, and give, the maximum cooperation in achieving high quality standards. Faster and more efficient means of inspection are constantly sought as it must always be remembered that inspection does not add quality to a product—only cost.

---

*Prepared for the U.K. vendors to a major American company.

CHAPTER 5

Purchase Inspection Procedures

The primary purpose of receiving inspection is to control the quality of purchased items which are accepted by the Company. This control of quality is exercised through the examining and sentencing of batches of product received from vendors. A perfect acceptance program would ensure the acceptance of all nondefective items and the rejection of all defective items. The program should also encourage and assist vendors in the improvement of the quality of their product. The program should be easy to administer and economical to operate. These two ideals are difficult to achieve in practice.

Decisions about the disposal of items of product must be based on an estimate of the quality of the product. Such an estimate of quality is always based on inspection of the product. The labor, equipment, and time required for this inspection are all costly, and many organizations have devoted much effort in investigating means of reducing those costs. The most useful and widely accepted result of those efforts has been the development of sampling inspection techniques based on the theory of probabilities. The validity of sampling inspection relies on the fact that quality may be determined by examining some, but not all, of the product submitted. Determination of the number of items to be inspected is based on a compromise between the high cost of inspecting all items and the cost that must result when defective material is stocked. There are four generally used methods of inspection: 100% inspection, percentage inspection, attribute sampling inspection, and variables sampling inspection.

100% Inspection. Sometimes called screening, 100% inspection is often resorted to when it is considered to be necessary to remove all defective items from a batch of work. In theory, at least, this method should be the only one to ensure that all defectives are removed from a batch. In practice, a false sense of security is engendered, except in those cases in which fully automatic inspection machines are used. In other cases the self-deception varies considerably. In any case, screening is invariably expensive in personnel and time. Also, by itself, it is not very likely to help in the improvement of future quality for the purchaser. Only defective parts are returned to the vendor. All nondefective parts are accepted. The vendor has thus, from this

VENDOR SELECTION AND APPROVAL

point of view, little or no incentive to improve product quality or to sort out defective items. In effect, the purchaser is doing part of the vendor's work, as it is normally part of the vendor's contractual responsibility to supply satisfactory product.

Percentage inspection. This type of inspection is also commonly used but like 100% inspection, gives a false impression of security. Usually, the procedure is to take a fixed percentage, often 10% of the batch of parts and sentence the batch according to the results of inspection of the sample. The great fault of this system is that the chance of accepting an inspection lot of any given quality varies much too widely with inspection lot size. For small inspection lots this leads to too little inspection for adequate protection, while for large inspection lots the expense of inspection is unjustifiably high. As soon as it is realized that in sampling inspection, the protection derived from inspection depends very largely on the total number of items inspected from an inspection lot rather than the percentage of the inspection lot inspected, interest in any fixed percentage dies.

Attributes and variables. In lot-by-lot sampling only some of the submitted items of product are inspected and complete lots, or batches, are accepted or rejected as whole. No sampling plan can ensure complete separation of defectives from nondefectives among submitted items, for not all the items are inspected. Accepted lots may contain some defective items and, conversely, rejected lots may contain nondefective items. Moreover, sometimes a low-quality inspection lot will be accepted and sometimes a high-quality inspection lot will be rejected. It has to be remembered that, while good inspection sampling plans prevent these errors from happening frequently, samples obey the laws of chance and will, therefore, occasionally be of a quality that is far out of line with the quality of the inspection lot from which they were drawn. When these risks are balanced against the considerable cost saving, sampling inspection will generally be preferred to screening.

The acceptance or rejection of an entire inspection lot on the basis of the quality of a sample drawn from that inspection lot provides a powerful incentive for the vendor to improve the quality of his product. If large inspection lots, even, are occasionally rejected, the supplier may find much of the profit eaten up by screening, reworking, and scrap. The way to reduce the number of lots rejected, and so reduce those costs, is to submit a product of higher quality.

Lot-by-lot sampling can be carried out either by attributes or by variables. The difference between the two is roughly as follows. In attributes sampling an item is merely good or bad and the sentence on the whole lot is based on the number of defectives in the sample. In variables sampling the particular feature being checked is measured directly and the sentence is based on a series of simple calculations.

Attributes plans are simpler but call for more items to be inspected for a given degree of protection than do variables plans. The latter, on the other hand, while administratively more difficult, provide much more information about the product. Both types of plan can be operated with the selection of single or multiple samples, but as far as the Company is concerned, only single sampling plans will be considered, although attributes or variables may be used.

It may be inevitable that certain parts will always require maximum inspection because of the need to remove all defectives. It is our intention that, with those exceptions, we begin using sampling inspection plans on a selected range of parts at an early date and gradually extend the range thereafter.

Effect on Vendors

During the operation of sampling inspection programs it will inevitably happen from time to time that batches of parts will be rejected. This is no different, generally, from what already happens. We are, however, much more sure of the general quality level of the batches of parts which we accept and which we reject.

We feel that over a period, and with mutual cooperation, the operation of a scheme on sampling inspection will result in advantages both ways. The principal advantage to be derived from this method of inspection is that it permits inspection costs to be reduced without in any way sacrificing quality.

Looking to the Future

As a very long term possibility, it may eventually be possible under this scheme for batches of parts to be accepted by the Company on

VENDOR SELECTION AND APPROVAL 59

the basis of certified results of a sample inspection carried out by the vendor.

The Company,

Quality Control Division

APPENDIX 2: ASSESSMENT, SELECTION, AND CONTROL OF SUBCONTRACTORS

1. **Purpose of the Procedure.** To describe the methods that will be used to assess, select, and control subcontractors with whom the Company will place orders for material, parts, or services or any combination of the three. To describe, also, the system of categories used by the Company to classify its subcontractors and to define the terms used in connection with it.

It should be noted that in this procedure, so far, the meaning which should be given to the term "subcontractor" is that meaning which is used in the British Ministry of Defence Defence Standards, that is, a company that supplies to another company. From the stage in the procedure when the categories are defined, only the Company meanings will be used.

2. **Scope.** This procedure applies to the activities of the supplier quality assurance engineering section of the quality assurance department and also to the purchasing department, which will, to some extent, be constrained in their freedom in placing orders by the requirements of the procedure. There may be a marginal involvement, also, of the goods inwards inspection area.

3. **Related and Relevant Procedures and Other Documents**

   Def. Stan. 05-21, paragraphs 210a, 205, 210c [see Appendix A]
   Def. Stan. 05-26, [see Appendix A]
   Def. Stan. 131A, [see Appendix A]
   03-701-001   Requirements for Subcontractor Records
   03-702-002   Evaluating a Subcontractor
   03-711-001   Requirements for Subcontractor Gauge Control and Calibration
   03-713-001   Release of Company material to Suppliers

4. Responsibilities

4.1. The prime responsibility for carrying out the requirements of this procedure belongs to the quality assurance supplier control (QASC) section. They also have the responsibility for guiding the purchasing department, which has a secondary responsibility.

4.2. The purchasing department has the responsibility for ensuring that they do not place orders with unapproved subcontractors. Alternatively, if there is a need to place an order with such an unapproved subcontractor, they must then ensure that a request is made to the QASC section to have the supplier assessed and approved, provided, of course, that the other requirements of this procedure are met.

4.3. Through the subcontract controller the purchasing department also has the responsibility for seeking out possible vendors, who are given a first screen. If this is satisfactory, the vendor is passed to the QASC section for complete action according to this procedure.

5. Operation of the Procedure

5.1. *Classifying Subcontractors*

5.1.1. To meet the varying and specialized needs of the Company, it is necessary that subcontractors be divided into a number of categories according to a predetermined grouping. This grouping, and the meanings given to each category, are described in the following paragraphs.

N.B. From this point on, the term "subcontractor" will be used only in the Company meaning.

5.2. *Categories and Definitions Thereof*

5.2.1. *Suppliers*

5.2.1.1. Suppliers are companies that supply items or material which are either proprietary or standard or stock items or material.

5.2.2. *Intermediary Subcontractors*

5.2.2.1. Intermediary subcontractors are companies that can, and do, carry out machining and/or processing operations on Company-manufactured items or parts at an intermediate stage in the sequence of manufacture by the Company.

VENDOR SELECTION AND APPROVAL 61

5.2.3. *Subcontractors*
5.2.3.1. A subcontractor is a company that will completely machine components from material supplied to them by the Company.

5.2.4. *Prime Contractors*
5.2.4.1. Prime contractors are companies that accept orders from the Company for the total manufacture of product to Company requirements. This includes all aspects from the supply of raw material to final testing of the completed product.

5.3. *Assessment of Companies That May Be Required to Supply to the Company*
5.3.1. Whenever the purchasing department has a need to place an order with a company that is not on the Company list of approved firms, they will take the following actions:

5.3.1.1. The buyer must complete a "Request for Approval of an Unapproved Firm" form (P1068) and send it to the QASC section. The form, apart from identification information about the firm concerned, must indicate the type of service that it is proposed would be ordered from the firm if it becomes approved.

5.3.1.2. A QASC engineer will assess and evaluate the firm in accordance with procedure 03-702-002. "Evaluating an Unapproved Firm."

5.3.1.2.1. In the special case in which the firm is being considered for processing work (e.g., heat treatment, plating, and similar special processes), the QASC engineer must call in the services of the chief metallurgist to approve the firm's processing practices, controls, records, etc. Approval of a firm for such processes will only be given on the personal, written, authority of the chief metallurgist, countersigned by the quality manager, and then, only for a single process on a single part.

5.3.1.3. In general, the information that will be gathered will include items under the following headings:

5.3.13.1. Physical manufacturing facilities, process capabilities, and state of maintenance.

5.3.1.3.2. The measurement and test facilities in the production, inspection, and technical departments.

5.3.1.3.3. The quality control manual and procedures for handling various quality activities (e.g., gauge control, deviation procedure, record systems, use of control charts, etc.).

5.3.1.3.4. The evident skills of the firm in quality matters (experience in production manufacture).

5.3.1.3.5. The apparent state of quality mindedness.

5.3.2. If the results of the evaluation are favorable, the firm will be categorized in one of the four categories in paragraph 5.1.1, and subparagraphs thereof, allocated a Company approval number, and the details recorded in the appropriate section of the Company's approved firms list. N.B. The Company approved firms list is in four sections, corresponding to the four categories already mentioned.

5.3.3. If the results of the evaluation are not favorable, the QASC engineer will enter full details in the lower section of the form originally raised by the buyer (see paragraph 5.3.1.1). No further action will be taken with regard to that firm.

5.4. *Orders*

5.4.1. Copies of every production order placed with firms on the Company approved firms list will be sent by the purchasing department to the QASC section.

5.4.2. These copy orders will be examined to ensure that the information on the order is correct and that the description of the material ordered is clear and that, as applicable, the following information has been included:

5.4.2.1. The type, class, style, grade, or other precise identification.

5.4.2.2. The title or other positive identification, and applicable issue, of any relevant specification or other drawing, process or inspection requirements, or other relevant technical information.

5.5. *Preproduction Evaluation*

5.5.1 Firms recently added to the Company's approved firms list may be required to submit preproduction samples for evaluation prior to full authorization for normal production of an order. It may also be that, as a result of review of records or for other reasons as may be agreed by the buyer and the QASC engineer, other firms will be required to submit preproduction samples. Handling of preproduction samples is dealt with in the following manner.

VENDOR SELECTION AND APPROVAL

5.5.1.1. The Company QASC engineer will raise a three-part set of sample inspection report forms on which the supplier will record his inspection results for the preproduction samples. The distribution of these forms is in accordance with the procedure for sample inspection report forms. (Additional sets of this form will be made available to the firm concerned if there is not room on the original form for all the details of the inspection of the samples.)

5.5.1.2. The relevant purchase order will contain details such as the number of items to be produced as preproduction samples and the general handling of the samples and paperwork. In any event the samples and paperwork will be handled internally in accordance with the procedure referred to in the preceding paragraph.

5.5.2. When the forms and the samples have been received by the Company and examined, and the result is acceptable, the form will be so endorsed and the firm will be instructed to proceed with full production.

5.5.2.1. In parallel with the action in paragraph 5.5.2, the QASC engineer will arrange for a bought-out inspection report form to be raised with details of the inspection carried out by the Company as confirmation of the results obtained. It will be distributed as normal for that form.

5.5.3. Should the results of the examination of preproduction samples not be acceptable, the QASC engineer will resolve the problem in consultation with the firm, one way or the other. If the result is favorable to the firm action will be taken as in paragraph 5.5.2. If the result is not favorable, the firm will be instructed to produce a fresh set of samples.

5.6. *Establishment of Quality Standards*

5.6.1. When large orders for bought-out components are placed by the Company, the firm with whom the order has been placed will be required to inspect shipments on a sample inspection basis prior to dispatch to the Company.

5.6.2. Acceptable quality levels (AQLs) will be directly related to the product concerned and will be determined by the QASC engineer. The product class and function will be taken into account in setting the acceptable quality levels with respect to the following factors:

5.6.2.1. The dimensional accuracies that are necessary because of the criticality of the part.

5.6.2.2. Functional sensitivity related to critical dimensions and associated costs.

5.6.3. *Sampling Plans*

5.6.3.1. All sampling plans used will be in accordance with DEF 131A [see Appendix A] and will encompass the following features:

5.6.3.1.1. The required levels of inspection.

5.6.3.1.2. The sample size.

5.6.3.1.3. The type of inspection to be used (i.e., single sampling and Normal inspection).

5.6.3.1.4. Random sample selection based on the tables in DG7A (a guide to the use of DEF 131A) [see Appendix A] when it is economically feasible to do so.

5.6.3.2. As suppliers may use other forms of sampling plans, adjustments to obtain estimated percent defective will be made so that the suppliers' AQLs may be evaluated.

5.6.3.3. All sampling plans used within the Company and given to suppliers are approved by the quality assurance department of the Company.

5.6.4. For parts of especially high degrees of criticality the supplier may be requested to carry out 100% inspections. Such requests may be made by the QASC engineer subsequent to the placing of an order. However, in those instances when a 100% inspection was not a requirement of the original order, all subsequent instructions for such inspections will be confirmed by means of an order amendment.

5.7. *Inspection by the Supplier*

5.7.1. When requested by the Company, the supplier's inspection department will record the results of their inspections on the Company form P1056, "Inspection Control Plan." A copy of this form, duly completed, together with any relevant test certificates or release notes will be sent to the Company with each shipment of parts. The information on form P1056 will include item identification, the nature and number of observations made, the number and types of deviations that were found, and the quantities rejected and approved.

5.7.2. If form P1056 records any major or critical defect that the supplier wishes to have considered by the Company for ac-

VENDOR SELECTION AND APPROVAL 65

ceptance, the supplier must request such consideration of the defect(s) on the Company concession/production permit application form, ADMEL 0999/1.

5.7.2.1. On receipt of such an application the Company's engineering and quality departments will consider the application in detail. When a decision has been reached, the application form will be endorsed with the decision.

5.7.2.2. Regardless of the decision, a copy of the application form will be returned to the supplier. In the event that the application is rejected, the supplier will be required to dispose of the parts at his own plant. If the application is accepted, the supplier will be required to ship the parts to the Company with the paperwork endorsed with the serial number of the concession/production permit form and with brief details of the defect involved.

5.8. *Supplier Gauge Control*

5.8.1. Suppliers will be required to carry out a system of control over gauges and measuring equipment that will, among other things, ensure traceability to national standards. They will also be required to maintain a system of records for gauge control which makes it possible to determine the current calibration position of each item of gauging and measuring equipment used on parts being made against one of the Company's orders. Generally, the supplier's system of gauge control and calibration should meet the requirements of British Defence Standard 05-26.

5.8.2. Similarly, in those instances when the suppliers have been provided with Company gauges and/or measuring equipment on loan, they will be required to exercise the same degree of control over them as their exercise over their own gauges and measuring equipment.

5.8.3. The Company QASC will also, from time to time, arrange for spot checks to be made on any of the Company's gauging or measuring equipment that is on loan to a supplier.

5.9. *Inspection of Highly Critical Parts*

5.9.1. Although the QASC engineer will from time to time carry out audits on the inspection and quality operations at the suppliers' plants, the engineer will also, on occasion, carry out 100% inspections on certain highly critical parts. In those instances the engineer will complete form P1056, which will be sent to the Company with the shipment by the supplier. However, such additional inspections as may be carried out by the Company QASC engineer will in no way relieve the suppliers of their obligations to supply to specification.

5.10. *Control of Supplier-Submitted Quality*

5.10.1. In exercising control on supplier-submitted quality and in arranging for the necessary information feedback from the Company to the supplier, a number of activities take place at the Company, which are described in the following paragraphs.

5.10.1.1. *Lot-by-Lot Variations*

5.10.1.1.1. At the Company's goods inwards inspection (GII), form P1056 is completed on each batch of incoming parts and is used to determine variability on a lot-by-lot basis.

5.10.1.1.2. The supplier will, in all cases in which variability of significance is observed, be informed by the QASC engineer.

5.10.1.2. *Summary of Data*

5.10.1.2.1. *Vendor Summary*

5.10.1.2.1.1. For each part number, by supplier, a summary sheet, P1051, is prepared. On this is recorded minimal information about each shipment of that part from that supplier [i.e., date and quantity received, quantities rejected (if any) and batch reject percentages].

5.10.1.2.2. *Weekly Summary*

5.10.1.2.2.1. Information similar to that recorded on the supplier summary is recorded on form P1077. This form is distributed internally in the Company to the purchasing department to inform them of shipments received, accepted, and rejected during the week.

5.10.1.2.3. *Monthly Summary*

5.10.1.2.3.1. Each month, from the various forms referred to in earlier subparagraphs of 5.10.1, a monthly summary is prepared. This summary provides the following information.

5.10.1.2.3.2. Suppliers who, during the month, have had a total of 5% or more batches rejected.

5.10.1.2.3.3. The total number of suppliers during the

month and the number who had 5% or more batches rejected.

5.10.1.2.3.4. The total number of batches received during the month and the number of batches rejected during the month.

5.10.1.2.3.5. A monthly trend graph on a running basis, together with an aggregate average.

5.10.1.2.3.6. A listing of each individual component that has, during the month, had a reject rate of 5% or over, together with the name of the supplier, the shipment quantities and rejection rates, reasons for rejection, and the action(s) taken.

5.10.1.2.4. *Six-Monthly Summary*

5.10.1.2.4.1. Each six months a further summary is prepared which brings together the information in subparagraphs 5.10.1.2.3.2 and 3 and 4.

5.10.2. *Rework of Nonconforming Materiel*

5.10.2.1. In the event that, for reasons of urgency, it is considered that received nonconforming materiel could be reworked by the Company, a cost for this rework will be determined. This cost will be advised to the supplier by the Company's purchasing department at the same time as a formal request is made for the supplier's permission for the rework to be carried out by the Company and charged back to the supplier.

5.11. *Additional Information*

5.11.1. In the event that a supplier is required to purchase materiel for use in satisfaction of a Company order against a Ministry of Defence contract placed on the Company, the supplier is required to obtain and furnish to the Company two copies of the relevant release notes or test certificates for such materiel.

5.11.1.1. The requirement in the immediately preceding paragraph will in all cases be part of the conditions stated on the Company purchase order.

5.11.2. *Amendments or Modifications to Drawings*

5.11.2.1. In every case in which a supplier receives from the Company amended or modified drawings with respect to an order already placed with the supplier, the supplier must receive an amendment to that purchase order. If such an order amendment is not received, the supplier must request one from the Company before proceeding with any work necessitated by the amendment or modification to drawings.

5.11.3. Normally, any heat treatment that may be required for parts or other materiel being supplied to the Company by a supplier will be carried out by the Company. Any deviation from this requirement must be agreed to in advance and will be covered by a purchase order or an amendment to an existing purchase order (see also paragraph 5.3.1.2.1).

5.11.4. In cases in which a supplier is supplying machined or partly machined components or items to the Company, they may be required, at the discretion of the Company as expressed through the purchasing department, upon completion of their final inspection to record the results on a company form P1056 (inspection control plan), copies of which will be supplied by the Company. In these cases one copy of the completed form must accompany each shipment.

5.11.5. The Company QASC engineer shall have the right to make such inspections as are considered to be necessary at the supplier's plant on product that is being processed to Company orders. These inspections may be at intermediate stages of the manufacturing process or at final inspection. Any such inspection carried out by the Company QASC engineer in no way absolves the supplier from his responsibility to supply to the requirements of Company orders, drawings and/or specifications.

5.11.6. *Supplier Subcontracting*

5.11.6.1. Suppliers are not permitted to subcontract against any Company order upon them without the express permission of the Company. This is a condition that will form part of the general purchase order conditions.

5.11.6.2. In the event that a supplier considers it necessary to subcontract any of the work placed upon him in a Company order, he must first obtain permission from the Company purchasing department.

5.11.6.3. When a supplier has received permission from the Company to subcontract any of his commitment they must, if at all possible, place subcontract orders with companies already approved by the Company for the class of

VENDOR SELECTION AND APPROVAL

work that he wishes to place with them or, which hold the appropriate Ministry of Defence approval. In all cases the supplier will inform the Company of the names of the companies with which he proposes to place subcontract orders.

APPENDIX 3: REQUIREMENTS FOR SUBCONTRACTOR GAUGE CONTROL AND CALIBRATION

1. Purpose of the Procedure. To set out the basic requirements for the control and calibration of gauges and measuring equipment used by suppliers to the Company on work which they are doing for the company.

2. Scope. There is direct concern on the part of the quality assurance supplier control (QASC) engineer because of his relationship with the supplier, but there is also an involvement on the part of the purchasing department because of their general concern for suppliers and, also, on the part of the Company gauge control supervisor, who will be involved in any gauge checking and/or calibration.

3. Related and Relevant Procedures and Other Documents

Def. Stan. 05-21, paragraph 205 [see Appendix A]
Def. Stan. 05-26, paragraph 211 [see Appendix A]
01-601-001   Quality Systems Audits
03-701-001   Requirements for Subcontractor Records
03-702-001   Assessment, Selection, and Control of Subcontractors
03-702-002   Evaluating a Subcontractor
06-101-002   Retention and Control of Records

4. Responsibilities. Because of their strong links with suppliers, the purchasing department has the prime responsibility. However, as part of his task requires him to consider this point, the QASC engineer also has a direct responsibility. Together they must ensure that any systems for gauge control and calibration which a supplier employs are submitted for the approval of the supervisor of the Company's gauge room who is directly responsible for Company gauge control and calibration. To the extent that the supplier's work for the Company involves the use of gauges and measuring equipment belonging to him, any controls that he exercises on those gauges and measuring equipment are regarded as an extension of the Company's own system for gauge and calibration control.

5. Requirements of the Procedure

5.1. During his initial assessment of a proposed supplier, the QASC engineer must obtain details of the supplier's gauge control and calibration procedure which would be used on any gauges and measuring equipment of the supplier that they would use on work for the Company. The engineer must ensure, by consulting the Company gauge control supervisor, that those procedures are satisfactory and compatible with those in use by the Company.

5.2. The supplier will, of course, be informed of those requirements and that he will be expected to meet them.

5.3. During the lifetime of any work that the supplier is doing for the Company, the QASC engineer will, at suitable intervals, carry out audits on the supplier's gauge control and calibration procedures in a manner that is convenient and suitable, jointly, to the supplier and the Company.

5.3.1. The Company gauge control supervisor will ensure that suitable records are maintained by the supplier of any supplier gauge and calibration control checks, and those records will be regarded as Company records subject to the same controls and constraints as the rest of the Company records.

5.3.2. Any audit carried out by the QASC engineer on the operation of the supplier's gauge control and calibration procedure is, in effect, an audit of a part of the Company system. It must, therefore, be made in full accordance with the requirements of the audit control procedure, 01-601-001.

APPENDIX 4: NATO INSPECTION SYSTEM REQUIREMENTS FOR INDUSTRY (AQAP-4, ED. 2)

North Atlantic Treaty Organization

Military Agency for Standardization (MAS):

NATO Letter of Promulgation

June 1976

1. "Nato Inspection System Requirements for Industry" (short title AQAP-4) is a NATO Unclassified publication. Periodic accounting is not required.
2. AQAP-4 is effective NATO-wide on receipt.
3. It is permissible to copy or make extracts from this publication without the consent of the Authorising Agency.
4. It is permissible to distribute copies of this publication to contractors and suppliers and such distribution is encouraged.

For the military agency for standardization

(Signed) R. Lawson,
Maj. Gen., RNLA
Chairman, MAS

CHAPTER 5

Table of Contents

Chapter 1		General	Page No.*
Paragraph	101	Scope	1-1
	102	Applicability	1-1
	103	Use of Existing Quality Control System	1-1
	104	Applicable Documents	1-1
Chapter 2		Requirements	
Paragraph	201	General	2-1
	202	Review and Evaluation	2-1
	203	Documentation	2-1
	203(a)	Inspection Procedures	2-1
	203(b)	Records	2-2
	203(c)	Technical Data and Changes	2-2
	204	Inspection Equipment	2-2
	205	Inspection of Purchased Materiel or Services	2-2
	205(a)	Purchasing	2-2
	205(b)	Purchasing Data	2-2
	205(c)	Receiving Inspection	2-3
	205(d)	Verification of Purchased Materiel	2-3
	206	In-process Inspection	2-3
	207	Workmanship	2-3
	208	Materiel Handling	2-3
	209	Sampling Procedures	2-4
	210	Non-conformance	2-4
	211	Final Inspection	2-4
	212	Packing, Preservation, and Marking	2-4
	213	Corrective Action	2-4
	214	Alternative Inspection Procedures and Equipment	2-4
	215	Accommodation and Assistance	2-5

---

*The page numbers given are those in the original document, not those in this copy.

VENDOR SELECTION AND APPROVAL

Chapter 1: General

101 Scope. This document establishes requirements for a contractor's inspection system. It identifies the elements of a system to be established and maintained by the contractor for the purpose of ensuring that materiel and services conform to contract requirements. The system shall be satisfactory to the Authority designated in the contract or its authorised representative, herein referred to as the Quality Assurance Representative.

102 Applicability. This document applies to all materiel and services when referenced in a contract or purchase order. If any inconsistency exits between the contract requirements and this document, the contract requirements shall prevail.

103 Use of Existing Quality Control System. The requirements of this document are less extensive than those prescribed in NATO Allied Quality Assurance Publication "NATO Quality Control System Requirements for Industry" (AQAP-1). When a contractor is producing materiel under a programme for quality which meets the requirements of AQAP-1 he may utilize his AQAP-1 system provided the inspection elements of his system satisfy the requirements of this document and are adequate to meet the inspection needs of the contract.

104 Applicable Documents. The following NATO document forms part of this document to the extent specified herein:

> AQAP-6 NATO Measurement and Calibration System Requirements for Industry

The applicable issue shall be the issue in effect at the date of invitation for bids unless otherwise stated in the invitation.

Chapter 2: Requirements

201 General. The contractor shall establish and maintain an effective inspection system to ensure that only acceptable materiel or services are presented to the Quality Assurance Representative. The Contractor shall maintain a documented inspection system capable of producing objective evidence that materiel or services conform to contract requirements whether manufactured or processed by the contractor or procured from subcontractors. The contractor shall ensure that essential inspection requirements are determined and satisfied throughout all phases of manufacture.

The contractor shall appoint a representative, preferably independent of other functions, to be responsible for all inspection matters. He shall have the necessary authority to execute such responsibility to the satisfaction of the Quality Assurance Representative.

202 Review and Evaluation. The inspection system established in accordance with the provisions of this document shall be periodically and systematically reviewd by the contractor to ensure its effectiveness and is subject to evaluation by the Quality Assurance Representative who may disapprove the system or any of its elements.

203 Documentation.

(a) *Inspection Procedures.* Clear, complete and current written inspection and test procedures, shall be prepared for each inspection operation, including those relating the assessment of the adequacy of process controls. Exceptions will be allowed when such procedures form an integral part of the detailed fabrication or process control documents, or when lack of written procedures will not adversely affect product quality. The contractor shall identify to the Quality Assurance Representative on request those inspection operations for which no written procedures are provided. Criteria for approval or rejection shall be included in all written procedures. Inspection equipment to be used shall be identified and its uncertainty of measurement stated.

(b) *Records.* The contractor shall maintain records of all inspection performed to substantiate conformance to contract requirements. Records shall include, as appropriate, identification of the item, the lot or batch, the nature and number of observations made, the number and type of deficiencies found, the quantities approved and rejected, and the nature of any corrective action taken. Records shall be retained and made available on request.

(c) *Technical data and changes.* The contractor shall ensure that the latest applicable drawings and specifications required by the contract, as well as authorized changes thereto, are used for inspection.

VENDOR SELECTION AND APPROVAL 75

204 Inspection Equipment. The contractor shall provide, calibrate and maintain inspection, measuring and test devices suitable to demonstrate conformance of materiel and services with contract requirements. Equipment shall be used in a manner to ensure measurements whose uncertainty is known and is consistent with required measurement capability. As necessary such responsibility shall include the validity of sub-contractors' measurement systems. Calibration of standards and measuring equipment shall be in accordance with AQAP-6.

205 Inspection of Purchased Materiel or Services

(a) *Purchasing.* The contractor shall ensure that all purchased materiel or services conform to contract requirements. When evidence of conformance depends solely on inspection performed by the sub-contractor the contractor is responsible to ensure that such evidence is statisfactory. Records of such inspection shall form part of the required contractor records. All sub-contracts and referenced data shall be available for review by the Quality Assurance Representative.

(b) *Purchasing data.* The purchasing document shall contain a clear description of the material ordered including as applicable:

1. The type, class, style, grade or other precise identification.
2. The title or other positive identification and applicable issue of specifications, drawings, process or inspection requirements, or other relevant technical data.

(c) *Receiving inspection.* Incoming articles shall not be used or processed unless inspected or otherwise ascertained to conform with contract requirements.

(d) *Vertification of purchased materiel.* The Quality Assurance Representative reserves the right to verify at source that purchased materiel conforms with requirements. Verification by the Quality Assurance Representative shall not relieve the contractor of his responsibility to provide acceptable materiel nor shall it preclude subsequent rejection. When the Quality Assurance Representative requires verification at source, the contractor shall include in his sub-contracts a statement which stipulates such requirement. The contractor shall also include statements which prescribe the Quality Assurance Representative's rights and authority within the sub-contractor's premises. The wording of such clauses shall be as directed by the Quality Assurance Representative.

206 In-process inspection. The contractor shall, as a minimum, perform inspections during fabrication on all characteristics which cannot be inspected at a later stage and ensure that process controls are implemented and effective.

207 Workmanship. Unless standards are stipulated in the contract, the contractor shall establish criteria for acceptable workmanship through written standards or representative samples. The contractor's proposed standards shall be satisfactory to the Quality Assurance Representative.

208 Materiel handling

(a) The contractor shall maintain procedures which precisely indicate the inspection status of materiel or work. Suitable identification media for such purposes are stamps, tags, routing cards, move tickets, or other control devices.

(b) As necessary, adequate storage facilities shall be provided to segregate and protect materiel pending use or prior to shipment.

(c) The contractor shall maintain procedures for handling inspected materiel to prevent abuse, misuse, damage or deterioration.

209 Sampling procedures. Sampling procedures used by the contractor shall be as stated in the contract or shall be subject to agreement by the Quality Assurance Representative.

210 Non-conformance. The contractor shall maintain procedures that ensure the identification of all non-conforming materiels, parts, work or articles and prevent use, shipment or intermingling with conforming materiel or work. Repairs shall be in accordance with procedures acceptable to the Quality Assurance Representative.

211 Final inspection. The contractor shall perform all inspection on the finished product or service necessary to complete the evidence of full conformance with contract requirements. Procedures for final inspection shall ensure that inspections that should have been conducted at earlier stages have, in fact, been performed and that the data is acceptable.

VENDOR SELECTION AND APPROVAL

212 Packing, preservation and marking. The contractor shall inspect the packing, preservation and marking and the materials used in those processes to the extent necessary to ensure conformance to contract requirements.

213 Corrective action. The contractor shall take prompt action to correct conditions that have caused or could cause either non-conformance of materiel or services, or deviations from agreed procedures and practices, or from the requirements of this document.

214 Alternative inspection procedures and equipment. Alternative inspection procedures and inspection equipment to those specified in the contract and associated documents, or process control inspection which ensures the materiel is of the required quality, may be used by the contractor when such procedures, equipment or control inspections provide, as a minimum, equivalent assurance of quality. Prior to their use, the contractor shall describe them in a written proposal and shall demonstrate for the approval of the Quality Assurance Representative their effectiveness in substantiating materiel quality. In case of dispute the procedures and equipment specified in the contract shall prevail.

215 Accommodation and assistance. The contractor shall provide the Quality Assurance Representative with the accommodation and facilities required for the proper accomplishment of his work and shall provide the assistance or data required by the Quality Assurance Representative for verification, documentation or release of materiel. The Quality Assurance Representative shall have the right to access to any area of the contractor's or his subcontractors' premises where any part of the work is being performed. The Quality Assurance Representative shall be afforded unrestricted opportunity to verify conformance of the materiel with the contract requirements. The contractor shall make his inspection equipment available for reasonable use by the Quality Assurance Representative for verifcation purposes. Contractor personnel shall be made available for operation of such inspection equipment as required.

CHAPTER 5

APPENDIX 5: A SELF-DETERMINATION PROGRAM FOR THE ESTIMATION OF THE DEGREE OF CONFORMITY TO THE THREE SYSTEM LEVELS IN PARTS 1, 2, AND 3 OF BS 5750/1987 (or ISO9001/3)

This program is based very loosely on an early British Ministry of Defence format, but has been totally revised and improved and brought up to date to conform with the requirements in BS 5750. The 18 assessment sections are presented in an approximately logical progression through the total manufacturing cycle, beginning with company policy and finishing up with delivery and personnel. *NOTE: This program can be applied equally well to the ISO 9000 series of standards, to MIL Q9858A and similar national standards.* The questions themselves are based on the requirements in Parts 1 to 3, of BS 5750/ISO 9000 although the wording in the standard itself may vary to some degree. This is partly because in those instances in which a requirement is similar in two or three of the parts, the wording that appears in the standard is not always the same. This has been explained as being due to a desire to "create differing levels of emphasis." Whether or not this was actually necessary is immaterial for the purposes of this program. A uniform wording has been used in all cases in which a requirement applies to more than one level.

Those who wish to try out this program should consider the questions with care. It has to be remembered that they are based upon requirements that are intended to apply to all sizes of company. This may well mean that, when they are being considered form the point of view of a small compnay, they will look rather formidable. With thought, however, it will be realized that they are not quite as bad as at first sight seems to be the case. This is borne out by the fact that many small companies have sought, and been granted, approvals at all three levels.

That said, it should not really be necessary to draw attention to the fact that before seeking and/or undergoing an assessment, a company should carefully consider its own organization with regard to the actual requirements for *the level in which they have an interest.* Yet there are companies that do not do so and are later surprized when approval is denied.

Completing this self-determination program provides a good relatively simple, and painless way to determine if the existing

VENDOR SELECTION AND APPROVAL 79

organization is likely to be approved after a formal assessment. There is always a certain amount of "flexibility" or interpretation in deciding if a particular situation meets the requirement. Excellent guidance is given in Parts 4, 5 and 6, as may be appropriate, of BS 5750 ("old version"). These parts are intended to provide guidance for assessors and, indeed, to companies seeking to prepare themselves for an assessment. With those points in mind, the respondent should be able to begin the task of working through the program.

Of course, respondents will not all be interested in the same level. Some will be interested in level 1. Some in level 2 and the remainder in level 3, which is the lowest, or simplest. Earlier analyses have shown that approximately 15% of companies have gained approval in each of levels 1 and 3 and the remainder in level 2.

To ease the task of the respondent even more, the program indicates the levels to which each particular requirement applies. It also identifies the relevant paragraphs in the three parts of BS 5750 from which the requirement comes. For example, in the section dealing with final inspection and test, question 10.1 asks: "Is there formal inspection and/or test to verify conformance to requirements?" Following the question the relevant paragraphs in BS 5750 are identified (e.g., in Part 1—4.10.2 and 4.10.3; in Part 2—4.9.2; 4.9.3; in Part 3—4.5). In the third of the three column to the right of the program sheets appear the figures 1 and/or 2 and/or 3, according to the part(s) of BS 5750 to which the requirement applies.

Look a little more closely at this question. The significant word is "formal." Of course, there will be "final insepection," but will it be formal? Do the inspectors have drawings and acceptance specifications as necessary? Do they have any written guidance to the performance of their tasks? Then, thought needs to be given to the relationship between this question and the preceding questions in Section 9. They all deal with inspection equipment and its calibration. Similar relationships will be found in a number of other instances. It should also be remembered that more will be expected with regard to question 10.1 of a company seeking level 1 approval than might be expected of a company seeking level 3 approval.

This program can not be "zipped through" in a half-hour or so in a real situation even if that were possible in an exercise. One must carefully consider all the questions before any attempt is made to answer "yes" or "no" to any of them. Answers should be given only after a detailed review of the actual situation on the shop floor or wherever. Depending on the size of the company and the level of approval being sought, it could be expected that proper completion of the program could take from less that 1 day to 3 days, or even more, if full justice is to be done to it, and if maximum benefit is to be gained. This may seem to be a lot of time, but if one considers the problems that would be faced, to say nothing of the extra expense and the damage to the company's image, if approval is denied after a formal assessment, it can be seen to be a very worthwhile exercise.

There are 71 questions in all in the program, all of which apply, of course, to level 1, 61 of them are applicable to level 2, and only 23 apply to level 3. (Note that question 7.1 is a multiple one) When completing the answer columns the respondent should place an "x" in the appropriate column opposite each question. When the question does not apply it is convenient, for later reference, to write "n.a." in the "YES" column.

Upon completing the task, the respondent will have a number of marks in the "yes" column and probably some in the "no" column. Although approvals have been given when one or two requirements have not been fully met, one should not be fooled by this "generosity" on the part of the assessment team. For one thing, they will be minor deviations of a kind that can easily be corrected. In this situation the assessors may well award an approval with the request that those few minor deviations be corrected immediately. Alternatively, the approval may be qualified, or even withheld, on the basis that the deviations must be corrected by an agreed upon date. A later visit would be made to confirm that the corrective action had taken place. (Note the difference between the two words "request" and "must.").

If however, any deviations are considered to be important, or if there are a number of them, there will be no approval. A formal request will be made for corrective actions to be applied, with the question of approval being dependent upon confirmation of the application of the corrective actions on a later assessment visit.

If the company has not done its homework—for example, by not carrying out this or a similar program—the consequences could be serious. It might be asked to carry out a complete review and

VENDOR SELECTION AND APPROVAL

up-grading of its organization before any further consideration could be given to approval.

It will be clear that judgment has to be exercised when considering the possibilities of gaining approval or not. But it is also stressed that the results of carrying out this program should be used to determine what has to be done to ensure that the answers are "yes." It really is not advisable to seek approval knowing that, even in, perhaps, one or two "minor" aspects, the requirements have not been met. The company may be lucky, but on the other hand, it might not. It is far better to be satisfied that all the requirements have been met and then secure the desired approval the first time. It is so much better for the image of the company to be able to say, perhaps; "Oh, we sailed through first time. We did our homework beforehand."

This program can also be used when a company wishes to seek approval to a higher level than the one currently held.

	YES	NO	Applicable LEVELS
SECTION 1. COMPANY QUALITY POLICY			
1.1 Are the company policy for quality, and the operational procedures for controlling product quality, formally documented? 　　1: 4.1.1;　2: 4.1.2.1			1,2
SECTION 2. COMPANY ORGANIZATION			
2.1 Are management structure, authority, responsibilities, and tasks of individual staff members formally documented? 　　1: 4.1.2.1			1
2.2 Is the quality organization directly responsible to a person independent of other functions? 　　1: 4.1.2.3;　2: 4.1.2.3; 　　3: 4.1.2.3			1,2,3

	YES	NO	Applicable LEVELS

2.3   Is there a company organization chart which identifies the job functions that have responsibility for design, manufacturing, and quality, and their status?   1,2
    1:   4.1.2.1;   2:   4.1.2.1

2.4   Is there a director or senior manager with authority to resolve all matters directly relating to quality?   1,2,3
    1:   4.1.2.3;   2:   4.1.2.3;
    3:   4.1.2.3

SECTION 3. REVIEW AND ASSESSMENT

3.1   Is there any arrangement for reviewing, at intervals, all the functions that are directly or indirectly related to quality?   1,2
    1:   4.1.2.2;   2:   4.1.3

SECTION 4. DESIGN CONTROL

4.1   Are there written design guides and/or codes of practice for the designers?   1
    1:   4.4

4.2   Is there any formal arrangement for the evaluation of new materials, reliability engineering, etc.?   1
    1:   4.8

4.3   Is there any system for reviewing contract requirements?   1
    1:   4.3

VENDOR SELECTION AND APPROVAL 83

	YES	NO	Applicable LEVELS

4.4 Is there any formal arrangement for the design department to supply data to those other departments that need it to assist them in carrying out contract requirements?
   1: 4.4.2.2   1

4.5 Is there a procedure for proposing, approving, and implementing design changes?
   1: 4.4.6   1

4.6 Is there any formal control over tests and trials?
   1: 4.2   1

4.7 Is there any design control over suppliers?
   1: 4.8   1

SECTION 5. PURCHASING

5.1 Is there a formal procedure for selecting and approving suppliers?
   1: 4.6.2; 2: 4.5.2   1

5.2 Are suppliers informed of quality and inspection requirements?
   1: 4.6.3; 2: 4.5.3   1,2

5.3 Are the quality arrangements of suppliers assessed in any way?
   1: 4.6.2; 2: 4.5.2   1,2

5.4 Do purchase enquiries and orders fully detail quality and inspection requirements?
   1: 4.6.3; 2: 4.5.3   1,2

	YES	NO	Applicable LEVELS

5.5   Does the quality department have any opportunity to verify the correctness of quality information or requirements on inquiries and orders?   1,2
    1:   4.6.3;   2:   4.5.3

SECTION 6.   WORK INSTRUCTIONS

6.1   Are there work instructions, inspection procedures, etc., available for the guidance of personnel?   1,2
    1:   4.2; 4.9.1;   2:   4.2; 4.8.1

6.2   Do the above instructions, etc., specify the required quality standards?   1,2
    1:   4.2; 4.9.1;   2:   4.8.1

6.3   Is there any system to verify that the above instructions, etc., are regularly used?   1,2
    1:   4.1.3; 4.17;   2:   4.1.3

SECTION 7.   PRODUCTION PLANNING AND CONTROL

7.1   Are contract requirements examined to confirm the following (*before* production commences!)
    (a) Material availability   1,2
    (b) Purchased materiel lead
        times   1,2
    (c) Equipment availability   1,2
    (d) Inspection and test
        techniques   1,2

VENDOR SELECTION AND APPROVAL 85

	YES	NO	Applicable LEVELS
(e) Manufacturing/inspection compatibility			1,2
(f) Contract delivery times			1,2

1:   4.3; 4.17;   2:   4.3; 4.8.1

SECTION 8. MANUFACTURING CONTROL

8.1   Are all manufacturing and inspection operations clearly defined in writing?   1,2
    1:   4.9.1; 4.17;   2:   4.8.1

8.2   Is materiel identification maintained, where necessary, throughout the manufacturing process?   1,2,3
    1:   4.8;   2:   4.7;   3:   4.4

8.3   Are the various inspection facilities fully compatible with requirements?   1,2,3
    1:   4.2; 4.11;   2:   4.2;   3:   4.6

8.4   Is any special process equipment approved or certified in any way on a regular basis?   1,2
    1:   4.9.2;   2:   4.8.2

SECTION 9. INSPECTION EQUIPMENT

9.1   Are there suitable facilities for storage, handling, maintaining, and calibrating all inspection and test equipment?   1,2,3
    1:   4.2; 4.11;   2:   4.10;   3:   4.6

	YES	NO	Applicable LEVELS
9.2 Are there formal procedures for equipment calibration? 1: 4.11; 2: 4.10; 3: 4.6			1,2,3
9.3 Are equipment accuracies traceable to national standards? 1: 4.11; 2: 4.10			1,2
9.4 Are suppliers required to meet the requirements above? 1: 4.6.2; 4.16; 2: 4.5.2; 4.15			1,2

SECTION 10. FINAL INSPECTION AND TEST

	YES	NO	Applicable LEVELS
10.1 Is there formal inspection and/or test to verify conformance to requirements? 1: 4.10.2; 4.10.3; 2: 4.9.2; 4.9.3; 3: 4.5			1,2,3
10.2 Are inspection and test requirements specified in writing? 1: 4.2; 4.10.3; 2: 4.9.2; 4.9.3			1,2
10.3 Does the quality department verify and approve all final inspections and tests? 1: 4.10.3; 2: 4.9.3			1,2

SECTION 11. SAMPLING PROCEDURES

	YES	NO	Applicable LEVELS
11.1 Do sampling inspection plans conform to any recognized standards? 1: 4.20; 2: 4.18; 3: 4.12			1,2,3
11.2 If not, do the plans used contain full approval and rejection criteria? 1: 4.20; 2: 4.18; 3: 4.12			1,2,3

VENDOR SELECTION AND APPROVAL

	YES	NO	Applicable LEVELS
11.3 Are defects classified? 1: 4.14; 4.20; 2: 4.12; 4.13			1,2

SECTION 12. INSPECTION STATUS OF MATERIEL

12.1 Is there a formal system for identifying the inspection state of materiel? 　　　　　　　　　　　　1,2,3
    1: 4.12; 2: 4.11; 3: 4.7

12.2 Is an accepted method of inspector and inspected materiel identification in use? 　　　　　　　1,2,3
    1: 4.12; 4.16; 2: 4.11; 4.16;
    3: 4.7

SECTION 13. CONTROL OF NONCONFORMING MATERIEL

13.1 Is there an accepted method for identifying, segregating, and disposing of nonconforming materiel? 　1,2,3
    1: 4.13; 2: 4.12; 3: 4.8

13.2 Is there a controlled segregation area for nonconforming materiel? 　　　　　　　　　　　　1,2,3
    1: 4.13; 2: 4.12; 3: 4.8

13.3 Is there a formal procedure for reviewing, reworking, repairing, and control of nonconforming materiel? 1,2,3
    1: 4.13.1; 2: 4.12.1; 4.13;
    3: 4.8

	YES	NO	Applicable LEVELS

SECTION 14. CORRECTIVE ACTIONS

14.1　Is there any formal system for prompt detection of nonconformance in materiel quality and applying suitable corrective actions?
　　1:　4.14;　2:　4.13   1,2

14.2　Is there any regular analysis of nonconforming materiel and records to detect and eliminate potential causes?
　　1:　4.14;　2:　4.13   1,2

14.3　Is any effort made to ensure that corrective actions are effective?
　　1:　4.14;　2:　4.13   1,2

14.4　Are any of the foregoing expected of, or applied to, suppliers?
　　1:　4.6.2; 4.16;　2:　4.5.2; 4.15   1,2

SECTION 15. RECORDS

15.1　Is there any system for recording, regularly, details of inspection and general quality activities?
　　1:　4.16;　2:　4.15;　3:　4.10   1,2,3

15.2　Is there any system for ensuring, regularly, that the records are accurate and maintained as required?
　　1:　4.5.1; 4.16;　2:　4.1.3; 4.15;
　　3:　4.1.3   1,2,3

VENDOR SELECTION AND APPROVAL

	YES	NO	Applicable LEVELS

15.3 Are the records in sufficient detail that full information is available for materiel, including quantities, inspection decision and disposition, etc.?   1,2
    1: 4.5.1; 4.16; 2: 4.4; 4.15

15.4 Is the foregoing expected of suppliers?   1,2
    1: 4.6.2; 2: 4.5.2; 4.15

SECTION 16. DOCUMENTATION AND CHANGE CONTROL

16.1 Is a complete list of documentation, including drawings and forms, that is used in contracts available for use?   1,2
    1: 4.5.2; 2: 4.4.1; 4.4.2

16.2 Are there suitable arrangements for ensuring that correct issues of all documents are at their points of use?   1,2
    1: 4.5.1; 2: 4.4.1; 4.4.12

16.3 Is there clear identification of responsibilities for the issue and revision of drawings, instructions, manuals, and other documents?   1,2
    1: 4.2; 4.4.6; 4.16; 2: 4.2; 4.4

SECTION 17. HANDLING, STORAGE & DELIVERY

17.1 Are there formal procedures that describe the methods for the protection of the quality of materiel

	YES	NO	Applicable LEVELS
during manufacture, inspection, and delivery? 1: 4.15.1/4/5; 2: 4.14.1; 3: 4.9			1,2,3
17.2 Are there suitable and segregated storage facilities as may be necessary, at all stages from receipt to final delivery? 1: 4.15.3; 2: 4.14.3; 3: 4.9			1,2,3
17.3 Do the storage methods in use take account of any special requirements? 1: 4.15.3; 2: 4.14.3; 3: 4.9			1,2,3
17.4 Is access to storage facilities restricted to authorized personnel, only? 1: 4.15.3; 2: 4.14.3; 3: 4.9			1,2,3
17.5 Are there separate storage facilities for customer-issued materiel? 1: 4.7; 2: 4.6			1,2
17.6 Is the stores issue system such as to prevent the issue of incorrect materiel and also, to ensure "first in, first out"? 1: 4.15.3; 2: 4.14.3			1,2
17.7 Is accepted materiel in storage inspected periodically to detect possible deterioration? 1: 4.15.3; 2: 4.14.3			1,2

17.8 Is identification of materiel in storage good enough to permit traceability back to analysis, inspection,

VENDOR SELECTION AND APPROVAL 91

	YES	NO	Applicable LEVELS
and test reports and purchase orders when required? 1: 4.8; 2: 4.7			1,2
17.9 Is there any inspection on packaging and despatch? 1: 4.15.4; 4.15.5; 2: 4.14.4; 4.14.5			1,2
17.10 Are there any formal instructions for the preservation, marking, and packaging of materiel for despatch? 1: 4.15; 2: 4.14; 3: 4.9			1,2,3

SECTION 18. PERSONNEL

	YES	NO	Applicable LEVELS
18.1 Are job descriptions available that set out the skills and qualifications required of personnel? 1: 4.1.2.1; 4.18; 2: 4.17; 3: 4.11			1,2,3
18.2 Are there training programs to upgrade personnel skills where appropriate? 1: 4.18; 2: 4.17; 3: 4.11			1,2,3
18.3 Are there special certification requirements for special process operators, such as welders and NDT operators? 1: 4.9.2; 4.18			1
18.4 Are training and certification methods subjected to regular review to verify their adequacy? 1: 4.9.1; 4.9.2; 4.18; 2: 4.8.2; 4.17			1,2

CHAPTER 5

		Applicable
YES	NO	LEVELS

SECTION 19. OVERALL QUALITY LEVEL ESTIMATION

19.1 Having completed this program, which, if any, of the three levels of quality and inspection systems in BS 5750 (or its equivalents in other national or international standards) does the compiler consider that his or her company could expect to meet?

1 or 2 or 3?

Note: The paragraph numbers conform to those in the 1987 edition of BS 5750 and ISO 9001, 9002 and 9003.

6
When It Is Received

Like the preceding chapter, this one deals with two different scenarios. The first one is that of the old and familiar, well-established, incoming inspection section of the type that is still to be found in so many of the smaller companies around the country (and larger ones, too). I had my first opportunity to establish just such a "purchased parts inspection" section more than 30 years ago for an American company, with further opportunities in subsequent years. One of the important lessons that I learned was the value of maintaining a close working relationship with my buying colleagues. The second scenario is that of the modern bought-out quality audit (BOQA) group. Groups such as this function equally well under either of the two regimes described in Chapter 5. That is, they can be used whether or not one also makes use of quality system standards such as BS 5750, the AQAPs, or MIL-Q-9858A. But first, scenario 1.

Basically the "on receipt" inspection of purchased parts is an expensive but unfortunately, necessary evil. It is expensive because, if it is to be done at all, it must be done properly, and that is expensive in terms of inspection labor and equipment. In large and sometimes in not so large companies, one will often find that the section of the quality/inspection department that deals with the on-receipt inspection of incoming material is large. This is an evil because the fact that it has to be done at all indicates that suppliers have not done their job properly—giving the customer what has been ordered in accordance with the specifications. Perhaps it should be pointed out here that most companies suffer in this respect, although some more than others.

Like so many other things, the extent of an on-receipt inspection facility depends on several factors: on what is being bought, from whom it is being bought, in what volume it is being bought, and on the technical complexity of that which is being bought. If it is done at all, it should be done properly, which means that all necessary gauges, measuring and test equipment, and so on, must be available. The planning section of the quality department must study carefully the drawings and specifications for the parts concerned so that suitable acceptance inspection specifications can be prepared. These specifications must be provided for the inspectors so that they have all the relevant and necessary information. It should also be said that it is important to ensure that the inspectors do actually use the latest issues of the relevant drawings and specifications.

Like everyone else, inspectors are fallible. When they have inspected the same part many times, they often tend to rely on their memories to save themselves the trouble of consulting the necessary drawings and specifications. They believe that this will save time and that it can be done because they can accurately remember what is called for. Sometimes they can. But more often, they cannot, and experience has shown that it is on these occasions that the well-known "Murphy's Law" operates. That will be the occasion when there have been new issues of either or both the drawing and the specification following changes. This seems especially to be the case when there is a need for an urgent decision. Experience has shown that in cases when urgent decisions are demanded, one should indeed "make haste slowly."

Few sections of a quality department need to work as closely with the buying department as does the incoming inspection section. Each is of value to the other. Perhaps it is not too much to say that each needs the other, especially for essential information. Incoming inspection may or may not have copies of the orders for incoming materiel, although they ought to have them, but they can be helped with advance information from the buyers about the arrival of urgent shipments so that they can be dealt with promptly. Vendor technical queries will be dealt with via the buyers, and so on. In turn, incoming inspection will keep the buyer informed about the progress of the inspection on incoming shipments of particular interest, and of the results, of course.

Many buyers will recall occasions when they have strained every nerve to get a shipment into the factory only to find out some days later when the production manager got upset with them because of the nonappearance of the parts, that the shipment had been received, promptly inspected, and equally promptly rejected and returned to the supplier. But no one had thought to inform the buyer while there was still time for him or her to take immediate remedial action.

Some large companies operate a system whereby it is the responsibility of the buyer to get incoming parts right onto the assembly lines. This is a system that has much to commend it, although it can make life difficult for both inspectors and buyers (and the supplier, too, it must be said). But of course, the responsibility of the buyer was not just to get parts onto the line, but usable parts that conformed to the specification—"right" onto the lines. This means that when a shipment is rejected, an engineering decision may have to be made as to whether the parts could be used on a concessionary basis. [It should be noted that the "just-in-time" (JIT) philosophy is beginning to be used by large companies.]

One sidelight on the operation of an incoming inspection section is the need for it be "secure"; that is, it needs to be fully enclosed and lockable after normal working hours. Without these precautions one may be sure that sometime or other, eager progress personnel will spirit away part or all of shipments before, during, or after inspection and without the knowledge of the inspection staff, and without the necessary paperwork having been cleared. In other words, security is of considerable importance here, as for so many other things.

It is not beyond the bounds of possibility that buyers, who have intimate connections with the running of an incoming inspection facility in a company that buys large quantities of parts to high technical standards, will have formed a pretty poor opinion of the general standard of quality that is supplied by industry in general. In one case in the United Kingdom, over a period of one year with an average of 1800 incoming shipments per month, ranging over 1600 parts and 200+ suppliers, no less than 18% of each month's shipments involved partial or entire rejection. Another 18% of each month's shipments had something wrong with them, although not enough to warrant rejection but involving the engineering review procedure. These figures suggest a very sorry state of affairs indeed. A large company in Camden, New Jersey, had practically identical results at about the same time.

Many reasons could be advanced for such a state of affairs, but two are likely to be predominant. The first is undoubtedly the existence of a weak quality organization at the suppliers. The second reason is that there is a very strong production control organization at the suppliers. Strong enough and much more concerned with meeting delivery dates at all costs regardless of any question of product quality conformance. There is no reasonable explanation for some of the cases experienced and one can only conclude that it is just a part of the general malaise that is affecting much of industry in the Western world as a whole.

But to return more directly to the role of the buyer in quality. Reference has been made to the importance of the transfer of information from incoming inspection to the buyer, and vice versa. That is, there should be an information feedback loop. Apart from the obvious reasons—passing back inspection information about received shipments—this feedback loop is necessary for other desirable reasons. One, which is somewhat intangible, may be equated loosely with one of the reasons for using feedback systems in audio amplifiers: that of stability. Stability in the relationship that should exist between the customer and the supplier is essential for a variety of reasons, not the least of which is that both parties will benefit from the realtionship. This subject is dealt with in much more detail in Chapter 8.

A side benefit of a feedback loop is that it allows the accounts department to clear invoices promptly for payment to suppliers. This will please the latter and it can also be used to provide an early warning signal to suppliers should there be a need to correct a deviation from specification before too many parts are incorrectly made and another shipment sent. The loop will keep buying personnel generally informed about the progress of the inspection of shipments. It is as important for the buyer to know about accepted shipments as about rejected ones. It should be a normal buying function to transmit quality information back to a supplier, although arrangements are often made that allow the two quality departments to contact each other directly, especially when urgent problems arise. But the buyer must be kept fully informed about what is taking place, especially of any acitivity that might, however remotely, affect delivery or price. It will often be the case that the solution ot a technical problem will affect either, or both, delivery and price.

Now for scenario 2: BOQA (*Note:* The BOQA organization described in this chapter was that of Company D.) The title "bought out quality audit" is comparatively new. BOQA is what to many might still be "goods inwards inspection" or something similar. In the apparently never-ending task of preventing defective materiel from passing through to assembly lines, BOQA is the last line of defence. This should not be taken to mean that it is intended to operate as a very intensive screen on a regular basis. Primarily, it should be regarded as an audit function to confirm that a vendor has done what he contracted to do. Deliver product that conforms to specification requirements. But if audit checks show that this is not the case, then BOQA can just as easily act as a fine screen to sort out the good from the bad. If, however, vendor quality assurance (VQA; discussed in detail in Chapter 9) and the supplier between them have done a good job, BOQA will be a purely audit function. Which is what it really ought to be.

At this point it is worth quoting the first two sentences from paragraph 210c, "Receiving Inspection," of AQAP-1, Edition 3.

> Incoming materiel shall not be used or processed unless and until inspected or otherwise found to conform to contract requirements.
>
> In determining the amount and nature of receiving inspection consideration shall be given to the controls exercised at source and the objective evidence provided.

The second of these sentences is the key to the whole basis of operation of a BOQA operation. It is made clear that just what it does "depends" on what the vendor does about controlling quality and the available evidence that the customer has to that effect.

A major task for BOQA, which helps it perhaps more than anything else to determine the extent and nature of the inspections that it will carry out on incoming materiel, is the operation of a vendor rating system. Vendor rating systems are discussed in Chapter 7. Suffice to say here that a well-planned vendor rating system is possibly the most objective way of producing "hard" evidence of the "goodness" of a vendor. Evidence provided by BOQA and without which the provision of such basic and valuable data as vendor rating would be much more difficult. It might be said that while VQA plans the war, BOQA wins the battles on its behalf. From the operational point of view BOQA should be part of VQA.

Although on the face of it a BOQA organization may seem to be quite straightforward, it is, in fact, quite complex if it is operating as it should. It has to handle shipment documentation, both in and out; carry out physical inspections or audits; create and maintain considerable records; operate a task planning function; liaise with good receiving, purchasing, material control, the laboratories, and other parts of the quality department; provide vendor rating data; and work closely with the rest of the VQA group and in particular with the quality surveyors. It should have its own manager, who will be responsible for all those tasks and who will be responsible to the VQA manager. With this outline picture of the organizational complexity of the BOQA operation, the various tasks can now be discussed on an individual basis—not necessarily, however, in the order in which they have just been given.

One of the most important tasks of BOQA is that of planning the extent and nature of the audit-inspections that are carried out. Earlier, a couple of well-known sayings in the quality world were mentioned: "inspection costs money" and "no inspection is good inspection." Whereas the first one is quite literally true, the second must be dealt with carefully, as, also literally, it could be that "no inspection is *bad* inspection," but it is a target that one should strive to achieve—one where the vendor's control over product quality is so good that all that BOQA needs to do is a minimal audit. The task of the BOQA planner has to take these two considerations into account in planning the inspection operations for the audit inspectors. The extent refers to the detail of the actual sampling plans that will be used. The nature refers to the specific features of an item that will be inspected by the auditor and the methods to be used. The intensity of the sampling will depend on whether the sampling plan is using "reduced," "normal," or "tightened" inspection at any particular stage. These terms will be explained later.

The extent of the inspection, in addition to varying according to the number of different features of an item that are being in-

WHEN IT IS RECEIVED

spected, can also be varied by adjusting the level of defective items that can be accepted by varying the acceptable quality level (AQL). This is an important topic that merits further discussion before dealing with the rest of the subject. It must first be remembered that one can either be an idealist or a realist in this world, and this also applies to the inspection task.

While in no way decrying the merits of philosophies such as zero defects (ZD) and, particularly, the much newer "parts per million" (PPM), and accepting that these are targets we should all be striving to achieve, one must be a realist. This means acceptance of the fact that humanity is what it is and that more often than not, things will not always be perfect. Rigid insistence that vendors maintain the zero-defect concept can have an unwelcome effect on costs. More about this later. For the moment one will simply say that the BOQA audit inspection planning must be realistic.

Although it would be nice to think that all our vendors would supply a product that is in complete conformance to specification, we know that this is not very likely. (See the quotation in Chapter 4 from the ASQC *Random Sample* newsletter.) After all, how many of us could swear that our companies never send out a defective product? Precious few, if any. And despite all that we hear, and that western consumers say and believe, about the nearly perfect performance of Japanese companies, there is hope yet for the West.

The audit inspection planner, when preparing the individual item inspection specifications for each item that is to be inspected, must first prepare a list of the various features on each of the items as they appear on the drawings and/or specifications. This done, the planner must then determine their relative importance. Contrary to what is often thought by the uninitiated, it would be most unusual if there were not differences in the importance of the various features. Frequently, these differences can be considerable. For example, if one takes a simple example such as a drinking glass of the type that one would find in the average bathroom, it would obviously be of more importance that the glass not have a hole in the bottom than that it contains exactly, say, one-third of a pint.

When the planner has identified all the necessary features and determined their relative importance, he or she must, as far as possible, then take account of the method of manufacture. Depending on this factor it will be found that in the case of some features, it will be most unlikely that there will be variations of any significance. In others there will be strong links between two or more features, because of the manufacturing method, so that if one changes, so will the others. The first being what one might call the "master," with the others as "servants." This is often the case with semiconductor components and other electronic components.

In those cases the planner can remove the "tied" features from the list of those to be inspected. Next the planner will look for those unlikely to be of significance for use and/or functioning. These can often be found. Sometimes they are called "thin air" features and may be required for manufacturing purposes only. In some industries many of these will be found. These features, too, the planner can discard. He or she will then be left with a reduced list of essential features which are of differing levels of consquence. Having got this far, the planner can differentiate between them by allocating different AQLs to the different levels of consequence (see BS 6001 or MIL-STD-105D).

The allocation of the AQLs is conveniently arranged according to a grading of the features that assesses their relative importance in the event of nonconformance. This relative importance is identified by use of the words "critical," "major," and "minor." There is a school of thought about this which considers that the use of such words "is illogical, the words themselves being incapable of precise definition and serving only to introduce ambiguity." But if one can judge by the numerous articles and papers that have appeared on the subject during the last several decades, it seems that most people in manufacturing do not hold this narrow view.

For these three words there are what seem to be perfectly adequate definitions available, of which the following are thought to be representative:

Critical: a defect that would cause a catastrophic product failure or would cause customer injury

Major: a defect that might cause a product failure or might cause customer injury

Minor: a defect that will not cause a product failure but would cause customer annoyance.

These are thought to be quite simple and straightforward. A possible amendment to "critical" might be to call it "safety critical." This takes into account the possibility of harm to the user and is a most important qualification in view of the fact that the laws of the European Economic Community countries are currently being changed to take account of the Directive on Consumer Safety. The definition could then be changed to the following:

Safety critical: a defect that would cause a catastrophic product failure and/or harm to a consumer.

Having set a group of defined grades for features that are to be inspected, and having classified them accordingly, it is now appropriate to consider the acceptable quality levels, AQLs, which should be allocated to them, taking, of course, the realistic as opposed to the idealistic attitude. The levels chosen will depend on the judgment of the planner, who must have at least reasonable experience of the type of product concerned and of the vendors concerned. The latter is very important. In addition to experience, the planner should also have available data from assembly and other inspection sources, including analysis of customer complaints.

In the first instance the planner may have to set levels "blind." But as soon as possible, he or she will have to take into account data from received shipments and if necessary, adjust the AQLs to take account of the later, factual data. Long experience would suggest that for the three grades a commonly chosen set of AQLs would be as follows:

Critical: Not higher than 1% AQL
Major: Not higher than 2½% AQL
Minor: Not higher than 4% AQL

The main effect that changing the AQLs is likely to have is to change the probablility of acceptance for the next shipment. A look at the single sampling tables of BS 6001 for normal inspection will show this. For example, in a sample of 125 pieces, the accept and reject numbers for an AQL of 1% are 3 and 4, respectively. That is, with 3 or fewer defectives, the sample is acceptable; with 4 or more defectives, the sample is rejected. For an AQL of 2½%, the accept and reject numbers are 7 and 8, respectively. MIL-STD-105D has identical tables.

The next thing that the BOQA planner must do is to prepare for the necessary differentiation to be made between the very good, the good, and the poor quality supplier. The differentiation is effected by means of adjustments to the amount of inspection that is carried out. Normally, three amounts, or levels, are used, although there are another four special ones that are rarely used. The three are called, in order of increasing amount of inspection, reduced, normal, and tightened. There is a very significant difference in the amount of inspection that is required by each of these levels. Taking normal inspection as the base level, which it usually is in any event, changing to reduced inspection will, in terms of the number of items in the sample actually inspected, bring down the amount of inspection by approximately 60%. The resultant cost gain will not be as large as this might suggest. This is because the time and costs involved in drawing a sample, recording the results, and disposing of the sample and shipment from which the sample was taken will not vary much regardless of the sample size. The change to reduced inspection will take place when sample inspection results signal the necessary "goodness" from vendors with very good conformance records. If, on the other hand, a change is made to tightened inspection, the average amount of inspection will increase by about 60%. Again, or rather to the contrary, the additional cost will be less than this amount, for the inverse of the reason given in the preceding paragraph and resulting from reduced inspection. This change will be made, of course, in the case of inspection results that lack the necessary "goodness" from vendors who have poor conformance records. There are, of course, recognized rules for making those changes in the inspection levels. They are called the switching rules, and full details about them will be found in the relevant standards, two of which were mentioned earlier.

There is a very important and little appreciated fact about these switching rules which should be mentioned, and emphasized. When using any recognized sampling plan, one can expect a certain degree of statistical protection. However, it is also the case that *if the switching rules are not used,* the degree of statistical protection that would otherwise be afforded is *reduced by up to 50%.* Sample personal surveys of mine strongly suggest that the majority of users of standard inspection sampling plans do not realize that these rules exist nor that it is important that they be used. Consequently, few sampling plan users apply them, and I must confess that I never did either until being informed about the consequences of not using them. So if full advantage is to be taken of the

WHEN IT IS RECEIVED

protection that standard sampling plans provide, *the switching rules must be used.*

The (the switching rules) are discussed, although not sufficiently emphasized, in both BS6001 and MIL-S-105D.

One other important task that the planner has to carry out is to determine the method of inspection that will be used for each feature. Is it to be measured directly by, for example, a micrometer or vernier caliper? Is it to be gauged by a standard gauge or a special gauge designed for that feature? Or, perhaps, is it a feature that will require laboratory examination? Taking into account the various factors of time, cost, difficulty of determination, and so on, the planner has to decide which method to use and then ensure that the necessary measuring equipment and/or gauges will be available.

The planner will have to work closely with the laboratories when considering the various inspections that will have to be carried out. The planner will decide personally on dimensional features. But for those that require the confirmation of a chemical or metallurgical or nondestructive form of test, the planner will have to consult the appropriate laboratory. Similar exotic features will include a variety of treatments, of which heat treatment is perhaps the most important, and finishing treatments, of which painting and plating are important. For all of these the planner must hold strictly to the requirements of the various laboratories.

To the planner, concerned as he or she is with the smooth flow of work through the BOQA area, laboratory requirements may seem to be burdensome; because of the additional sample that the audit inspector, for example, will have to select from the shipments for the laboratory tests, and because, perhaps, of the additional time that will be required for the laboratory tests, time which may mean that a shipment will have to be held for completion of time-taking tests when it has otherwise been cleared by the BOQA.

A further point that the planner may have to take into account is that the laboratory may not be able to fix beforehand a standard sample for a particular feature. The planner will have to insert into the specifications an instruction to the audit inspector to the effect that when the inspector begins inspection on a shipment, he or she will have to contact the appropriate laboratory to be told what their requirements are going to be.

From the work that has been described in the previous paragraphs, the BOQA planner will have prepared an acceptance specification for each of the items that are being bought in. The format for these specifications is not terribly important as long as it is one that provides all the necessary information for the audit inspector, is easy to read and understand, and is uniform for the entire series of specifications.

A comprehensive series of records will have to be developed and, as mentioned in an earlier chapter, the methods, styles, format, and contents will depend very much on whether written records are used or, alternatively, computer records. Whichever is the case, a lot of basic information will have to be recorded. The records will be linked with those of the goods receiving and purchasing department. There is likely to be interest from the material control department as well, if only to discover whether shipments have been accepted or rejected. Records are discussed in more detail in Chapter 13.

For our present discussion the records that will be of greatest importance will be the data about the results of the audit inspections on every received shipment. These will be used, as described in Chapter 7, for the compilation of vendor rating data. This will be invaluable as a guide for the purchasing department when the placing of future orders is under consideration. The data will also be a guide to the planner in his or her general considerations about the nature and extent of the audit inspections to be carried out. Furthermore, the vendor ratings will be a confirmation, or otherwise, to the VQA surveyors of the correctness of their own initial subjective gradings of vendors, gradings that may have to be replaced by the vendor ratings.

A speedy throughput of work is important for a BOQA area, and it is worth spending a good deal of time in planning the working area with this factor in mind; fixed workplaces for the audit inspectors, and fixed workplaces for particular major items or classes of items which may be associated with particular pieces of measuring equipment or gauges not readily portable. Separate areas for work awaiting examination, for work being dealt with, for work that has been cleared, and for work that has been quarantined pending a decision regarding disposition. Perhaps it is a statement of the obvious—although the obvious often needs to be stated—but a work area that looks cluttered up and untidy is not

WHEN IT IS RECEIVED

likely to be the most effective work area.

To sumarize what has been said in this chapter, it may be that the BOQA operation is a basic and essential part of VQA. It is also a complex one that requires careful planning and organization, and its successful functioning is vital to the VQA group in its overall task. Also, one must not forget that it is also a provider of information for many other parts of the company organization, not least of which is the purchasing department, with which it should liaise closely.

In three appendixes to this chapter, separate procedures have been included for the benefit of readers who want to delve more deeply into the practicalities of these situations. The first is a general procedure governing incoming inspection and the second deals with metallurgical and chemical inspection of incoming material. These are both for Company A. The third one is for control of incoming material and is for Company B. The first and last of this trio clearly show the nature of the differences that can be expected in the operating systems of average/sized and small companies. Again, small changes and deletions have been made to make sure that the identities of the companies are not revealed.

APPENDIX 1: GENERAL PROCEDURE GOVERNING INCOMING INSPECTION

1. Purpose of the Procedure. The procedure governs the general arrangements for the inspection of incoming material for production purposes. There are also exclusions and variations from the normal which will also be identified in this procedure.

2. Scope

2.1. The procedure covers the inspection. *in the goods inwards inspection (GII) area,* of the following range of items:

Sheet and bar
Forgings
Castings, including ferrous and nonferrous
Machined parts
Fasteners
Pressed parts
All packing material.

2.2 The following items are dealt with as may be instructed:

Jigs, fixtures, tools, or gauges
Consumable and maintenance stores items
Special parts and processing thereof
Other complex parts
Paint, chemicals, and dangerous items in general

3. Related and Relevant Procedures or Other Documents

05-203-001	Metallurgical and Chemical Inspection of Incoming Material.
05-202-002	Incoming Inspection Documentation and Records
05-202-003	Incoming Inspection of Complex Items
05-303-008	Incoming Inspection of Chemicals
01-601-001	Handling, Identification, and Segregation of Material
06-101-001	Control of Nonconforming Material
03-702-001	Subcontractor* Selection, Approval, and Control
05-202-004	Incoming Inspection of Special Parts and Processing

Def. Stan. 05-21, paragraphs 210a and 210c
Def. Stan. 131A [see Appendix A]

4. Responsibilities. Direct responsibilities according to this procedure are limited to the GII area and the metallurgical laboratory. There are, however, indirect responsibilities which extend to and from other quality assurance and inspection areas with regard to a small number of items which are referred to separately later.

5. Procedural Requirements

5.1. *Receipt of Material.* There is a single receiving channel through which material enters the factory: the goods inwards receiving area. However, material that is to be inspected by GII will, upon receipt and after issue of a goods inwards receiving note (GRN), proceed in one of seven paths, as follows:

5.1.1. Bar and sheet for special parts is delivered to the special parts division storage where it awaits inspection. All copies of

---

*"Subcontractor" here used in the Def. Stan. meaning.

WHEN IT IS RECEIVED

the GRN are delivered to GII as a signal that a shipment awaits inspection.

5.1.2. Any other shipment that is easily handled is delivered to the "awaiting inspection" shelves immediately outside the GII area and thereafter as in 5.1.1.

5.1.3. Large palletized shipments may be delivered directly into the GII area with the GRN copies.

5.1.4. Generally, large or heavy shipments stand in the enclosed "quarantine" area between the special parts division storage and the GII area and are dealt with thereafter as in 5.1.1.

5.1.5. Shipments of high-risk components for the special parts division may be delivered to that division's storage facilities and thereafter as in 5.1.1.

5.2. *Dimensional and/or Physical Inspection of Material*

5.2.1. In the case of material for which a supplier's test certificate or release note is required, inspection will not begin until the document has been received by the chief inspector and been passed to the GII area.

5.2.2. All sampling inspection is carried out in accordance with the requirements of Def. Stan. 131A and any other relevant sampling plans supplied by the parent company.

5.2.3.

5.2.3.1. When a shipment is to be inspected the supervisor passes the four copies of the GRN to the inspector assigned to inspect the shipment with the test certificate or release note when applicable.

5.2.3.2. The inspector will then collect the relevant drawing from the inspection record office (IRO) and, also there, check the particulars from the GRN against the relevant purchase order (PO), a copy of which is also available there.

5.2.3.2.1. When optical inspection is part of the inspection, the inspector draws the necessary master large-scale diagrams from the special parts division drawing library.

5.2.3.3. Except in the case of bar and sheet, the inspector will then also select from the GII files the relevant incoming inspection instruction (II), which contains the necessary detailed instructions for the inspector regarding the inspection task. The inspector is then ready to begin inspecting the shipment.

5.2.4.

5.2.4.1. In the case of bar and sheet the inspector will carry out a dimensional inspection at its location as indicated in 5.1.1 and 5.1.2.

5.2.4.2. In the case of shipments on the "waiting inspection" shelves or on pallets waiting in the GII area, the inspector will remove the shipment to his or her workplace and begin the inspection according to the requirements of the III.

5.2.4.3. In the case of heavy or large shipments located as indicated in paragraphs 5.1.3 and 5.1.4, the inspector will select samples from those locations, remove them to his or her workplace, and begin the inspection in accordance with the III.

5.2.4.4. High-risk shipments (see paragraph 5.1.5) will be inspected in the "safe" area, to which the inspector will take the necessary drawings, other required documents, and gauges and measuring instruments.

5.3. *When GII Inspection Results Are Acceptable*

5.3.1. *Further Examinations*

5.3.1.1. The inspector will refer to the metallurgical laboratory test plan summary and if the item is not on an "excluded" category will proceed as follows:

5.3.1.2. The inspector will take the top copy of the GRN and a copy of the test certificate or release note, as the case may be and if applicable, and send the pack of papers to the metallurgical laboratory as an implicit request for instructions regarding any metallurgical, chemical, and/or mechanical testing that the laboratory may wish to carry out.

5.3.1.3. Depending on the relevant past records, the metallurgical laboratory will determine the schedule of tests they wish to carry out and prepare three copies of the form "MSD Raw Material Testing," on which their requirements regarding samples and tests have been entered. The pack of papers, including the three copies of the MSD form, is returned to GII.

5.3.1.4. Upon receipt of the papers from the metallurgical laboratory, the inspector will take the required sample from the shipment, or arrange for it to be cut from bar or sheet in accordance with the instructions of the metallurgical laboratory, and send it back to the metallurgical laboratory with the pack of papers minus the test certificate or release note and the GRN notes. The last is filed in the GII "waiting" file to await the results from the metallurgical laboratory.

5.3.1.5. While awaiting the test results from the metallurgical laboratory, the inspector will prepare additional paperwork as follows:

5.3.1.5.1. A "Quality Report Diagnostic Chart" is prepared to record dimensional trends for high-volume batches and is filed in GII.

5.3.1.5.2. For all material other than bar and sheet (no additonal paperwork is raised for them) the inspector enters full details of the inspection, with results, on form P1086, "Inspection Control Plan."

5.3.2. When the metallurgical laboratory testing is satisfactory:

5.3.2.1. Upon completion of the testing, the results will be entered into an "MSM Results" form, which, with one copy of the original MSM form, is returned to GII.

5.3.2.2. The inspector will stamp "Accepted" and his or her own identification on all copies of the GRN. In the case of a shipment that is readily handleable and is within the GII area, the shipment and the yellow copy of the GRN are placed on the "Inspected Material" shelves immediately outside the entrance to the GII area. Later, the material will be moved into the storage facilities by storage personnel. The remaining copies of the completed GRNs are distributed to the appropriate departments.

5.3.2.3. For other shipments located as in the subparagraphs of 5.1, except for 5.1.2 and 5.1.3, the GRN yellow copy is placed with the material.

5.3.2.4. The remaining paperwork is filed in the GII files except that any test certificates or release notes are returned to the IRO which files them.

5.3.3. When the metallurgical laboratory testing is unsatisfactory:

5.3.3.1. In such a case, then, in addition, the "MSM Results" form and the "MSM Rejection Report" are also prepared by the metallurgical laboratory. On them are entered full details of the test results and the reason for actual rejection.

The two forms, together with the other forms as in the subparagraphs of 5.3.2, are returned to the GII area.

5.3.3.2. *Material Rejections*

5.3.3.2.1. The inspector prepares a "Bought-Out Inspection Report," on which are entered full details of the rejection. This form is sent to the IRO.

5.3.3.2.2. The inspector stamps "Rejected" in red ink on all copies of the GRN, and in the case of shipments as in paragraph 5.3.2.2, will place the shipment and a copy of the GRN together in the "Rejected Shipment" area adjacent to the GII to await return to the supplier or other appropriate disposition. The remaining paperwork is filed in the GII files or otherwise as in paragraph 5.3.2.4.

5.3.3.2.3. The IRO, from handwritten form R494, type a four-part "Bought-Out Inspection Report," which is distributed as follows:

First (white) copy to the supplier
Second (pink) copy to relevant buyer
Third (blue) copy to shipping
Fourth (yellow) copy to the IRO files

5.3.3.2.4. Upon receiving their copy the shipping department will make the necessary arrangements for the rejected shipment to be returned to the supplier.

5.3.4. Unsatisfactory GII results for material other than bar or sheet.

5.3.4.1. The inspector completes the first part of an "Incoming Material Quality Action Request" form. The material is placed in the "quarantine" area of GII to await sentencing. The form is sent, in sequence, to production control, engineering, and the planning section of the appropriate manufacturing division for a decision; for example:

Return to supplier as not acceptable.
Return to supplier for rework.
Rework in-house.
Accept on a concessionary basis.

5.3.4.2. When the decision has been made and entered into the form, with the necessary authorizing signatures and data, the form is returned, via the supplier quality assurance engineer for information, to GII for their action in disposing of the shipment according to the decision.

5.3.4.3. In the event that the decision is one of the first two referred to in paragraph 5.3.4.1, the actions taken are as in paragraphs 5.3.2.2.1 to 5.3.3.2.4 inclusive.

5.3.4.4. If the decision is either one of the last two mentioned in paragraph 5.3.4.1, the inspector will proceed as in paragraph 5.3 and relevant subparagraphs.

5.4. *Summary Data*

5.4.1. On a weekly basis GII will prepare a summary of the results of the inspections completed during the week. Copies will be distributed to the purchasing and production departments.

APPENDIX 2: METALLURGICAL AND CHEMICAL INSPECTION OF INCOMING MATERIAL

1. Purpose of Procedure. To set out the basis for determining the extent of metallurgical, chemical, and mechanical testing that is carried out on incoming material. To describe the system for reviewing the results obtained and use of the review in varying the extent of routine testing.

2. Scope of the Procedure. This procedure primarily affects the activities of the metallurgical department, although as it also concerns the work of the goods inwards inspection (GII) and to some extent determines the volume of its work, its scope includes the GII.

3. Related and Relevant Procedures or Other Documents.

Def. Stan. 05-21, paragraph 210c
05-202-001   General Procedure Governing GI Inspection
06-101-002   Retention and Disposal of Records
06-302-001   Control of Special Processes Carried Out by Suppliers to the Company

4. Responsibilities. The prime responsibilities within this procedure belong to the metallurgical laboratory, which must carry out all the work involved. However, there are minor responsibilities for the design and purchasing departments: the first named to ensure that through the normal channels of communication within the company, the laboratory is made aware of the availability of drawings and/or specifications for new materials or parts; the second named to ensure that the laboratory is always aware in advance of the identities of suppliers from whom materials, parts, or processing will be bought.

5. **Operation of the Procedure**

   5.1. *Determination of the Extent of Testing That Is to Be Carried Out*

   5.1.1. When the laboratory is informed of the existence of a new part or material and/or a new supplier, it will immediately request copies of the relevant drawings and/or specifications, under the authority of the chief metallurgist, from the special parts division library.

   5.1.2. Upon receipt of the drawings and/or specifications the chief metallurgist, or a staff member designated by him or her, will consider them carefully in association with the person's own experience and recorded knowledge of the capability of the supplier concerned.

   5.1.3. From this consideration the chief metallurgist or delegated staff member will determine a schedule of tests—metallurgical, chemical, and mechanical as necessary—which will be applied on a routine basis to incoming shipments of the new part or material from the new supplier.

   5.1.4. Complete details of the testing that is to be carried out, together with other relevant information (e.g., part number, drawing/specification numbers, supplier, and full specification details), will be entered into the raw material record card (form P1000) by the person referred to in paragraph 5.1.2. This person will also sign and date the card as authority for its subsequent use in reviewing the type and extent of the testing that is to be carried out and determination of variations in such type and extent of testing.

   5.2. *Review of Results*

   5.2.1. The reverse side of form P1000 is ruled off in a manner that permits the entry of the summarized results of the testing of successive shipments of that part or material from that supplier. This is done on a routine basis.

   5.2.2. After each routine entry has been made, form P1000 is examined by the chief metallurgist or delegated staff member (see paragraph 5.1.2). During this examination the cumulative results will be considered, and taking into account any other information that may be available since the previous review of the material, part, or supplier, a decision will be made as to whether any change should be made in either the test requirements or the extent of the testing.

5.3. *Action Following On from the Review*

5.3.1. If the decision made is "no change," no further action takes place.

5.3.2. If a change is to be made, the person concerned will complete a new form P1000 with the changed particulars. Summary reasons for the change must be entered. The new card will be used for routine entries as in paragraph 5.2.1, until it, too, is replaced or until a continuation card is needed.

5.3.3. Cards in use prior to a change, and cards in which all the useable space has been filled, should be filed for reference with the current card(s).

5.4. *Retention/Disposal of Records*

5.4.1. Completed cards should be retained for at least as long as may be required by the Ministry of Defence quality assurance representative. These requirements are detailed in procedure 06-101-002, "Retention and Disposal of Records."

APPENDIX 3: CONTROL OF INCOMING MATERIAL

1. Purpose of the Procedure. To ensure that production material that is received by the Company is not put into storage until its conformance to the specified requirements has been confirmed.

2. Scope. This procedure applies to the storage personnel, the inspection engineers, and the buyer.

3. Related and Relevant Procedures and Other Documents

- No. 5   Procurement
- No. 7   Stores Control
- No. 14   Function and Control of Inspectors
- No. 16   Records
- No. 17   Preservation and Packaging
- No. 18   Corrective Actions

4. **Responsibilities.** Responsibility for carrying out the requirements of this procedure rests primarily with the storage personnel, secondarily, with the inspection engineers, and lastly, with the buyer.

5. **Requirements of the Procedure**

   5.1. *Storage Personnel*

   5.1.1. As material for production purposes is received into the Company, the storage personnel will carry out the following tasks:

   5.1.2. The material will be checked for correct description and quantity against the delivery documents and the relevant copy order in the storage record system.

   5.1.3. The storage personnel will examine the material for any signs of transit or other damage—visual examination only—and note the results of this examination on the receiving documents.

   5.1.4. The storage personnel will then advise the relevant inspection engineer that material is waiting for inspection for technical conformance to specification requirements.

   5.1.5. When the inspection engineer has verified conformance, the storage personnel will place the material into stock and amend the storage records accordingly.

   5.2. *Inspection Engineer*

   5.2.1. When advised by the storage personnel that material for production purposes has been received and is ready for inspection and acceptance, the inspection engineer will make such inspections and/or examinations as considered to be required and/or necessary.

   5.2.2. If the results of the inspection are satisfactory, the inspection engineer will endorse the receiving documentation to that effect. This endorsement will be the authority for the storage personnel to accept the material into stock.

   5.2.3. If the results of the inspection are not satisfactory, the inspection engineer will take one or other of the following actions: (a) or (b).

   5.2.4. (a) The inspection engineer will authorize the shipping department to return the material to the supplier. He or she will endorse the receiving documents accordingly. The engineer will also advise the buyer and the accounts department.

WHEN IT IS RECEIVED

5.2.4.1. The buyer will ensure that the necessary documentation is prepared for the return of the material, including formal advice of rejection to the supplier, giving the reason for the rejection.

5.2.5. (b) The inspection engineer will consult with the project designer for whose project the material is required and/or the technical director of the company. Depending on the conclusions that are reached after these consultations, the material will be accepted or rejected.

5.2.5.1. If the material is to be rejected, the actions described in paragraphs 5.2.4 and 5.2.4.1 will be taken.

5.2.5.2. If the material is to be accepted, the actions already described in paragraph 5.2.2 will be taken.

5.2.5.3. When this material is accepted the inspection engineer will ensure that the circumstances are recorded on the receiving documents and that the buyer is also informed so that details can be entered into the file for the supplier concerned.

5.2.5.3.1. The buyer will ensure that the supplier is informed of the occurrence and requested to ensure that steps are taken to ensure that future deliveries conform to the requirements.

5.2.5.3.2. Included in the considerations taken into account under paragraph 5.2.5. may be included the possibility of corrective actions to be taken by the Company, including charging back to the supplier, after the supplier's agreement has been obtained.

7
Vendor Rating and Surveillance

In this chapter consideration is given, first, to a simple method of initial grading of vendors to provide a first guide for the BOQA planner, mentioned in Chapter 6, when he or she is preparing an audit specification. Second, a simple and practical system for vendor rating is described in detail. The value of a vendor rating system is illustrated by means of an actual case study, which demonstrated, beyond all doubt, the advantages of such a system.

As explained earlier, it is only in the simpler, questionnaire form of vendor assessment that gradings are usual. In the first instance, and until the gradings are confirmed or otherwise by actual performance, deciding on the grading to apply to a vendor is a subjective exercise. It is based on the opinion of the surveyor, which itself depends on the views that he or she formed about what he or she saw and was told during a visit to the vendor. At this time it is unlikely that there is any way in which the planner can form an objective assessment of the capabilites of the vendor's systems or personnel. There is no simple "scale of 10" against which the planner can relate the performance of the vendor in preventing nonconforming materiel from being shipped to a customer—to say nothing of any ability to prevent it from being made in the first instance!

The surveyor must use experience, instinct, and visual impressions. He or she must also remember that "nor all, that gilsters, gold" ("Ode on a distant prospect of Eton College" by Thomas Gray.) and that a "good front" may conceal poor performance. The reverse is also true: A poor-looking setup may turn out a high-grade performance. I remember a case during World War II, in which what can only be described as a pretty "tatty" small sheet metal shop turned out first-class product because the working owner, an elderly man, was a craftsman of the first order. However, the surveyor should remember these facts and that a grading then allocated must be regarded as temporary until the vendor has demonstrated his degree of conformance to requirements for his deliveries. Once this stage has been reached, the original gradings should be cancelled and forgotten. The subjective has been replaced by the objective; fiction by fact.

It must be said, however, that for the purposes of initial planning at the customer's incoming inspection or bought-out quality audit group, a grading is necessary. It provides that group with a measure to work to in deciding on the initial degree and level of audit inspection that will be carried out. The odds are that those vendors considered to be good by the surveyor will be good; that those considered by him to be not so good will be not so good. So now there is an easement of the task for the BOQA planners. Some thought can now be given to a system of grading to be used.

As long as gradings are determined, it is of little consequence what they are called. They could even, if one wished, be given names like those of the hurricanes that regularly sweep the American seaboard. However, it is more usual to give them numbers or letters. That will be considered later. In general, there is a case for three basic gradings. A fourth could be used, and some systems do provide for that. That is the "not acceptable" vendor. However, mature consideration should quickly decide that if a vendor is found to be not acceptable, it is simpler to say so than to call him a grade "x" vendor. In any case, the vendor can hardly be that, as he is not a vendor at all. This leaves the three already suggested.

   1. *The highest grade.* A vendor in this category is expected to achieve such a degree of conformance to specification requirements that little or no inspection should need to be carried out on received shipments; a supplier who, additionally, will also take due regard of any other requirements of a pro-

VENDOR RATING AND SURVEILLANCE

cedural nature that may be required by customers. It must be expected that only a minority of vendors will be considered eligible for this grade.
2. *The middle grade.* A vendor in this category will be expected to achieve such a degree of conformance to specification requirements that no more than a reasonable amount of inspection will have to be carried out on received shipments. His degree of conformance to other requirements of a procedural nature is likely to be of the same degree. It would be expected that a majority of vendors will be eligible for this grade.
3. *The lowest grade.* A vendor in this grade would be expected to achieve a low degree of conformance to specification requirements, so that significantly more inspection will be required. His degree of conformance to other requirements of a procedural nature is likely to be equally poor. It would be expected that only a minority of vendors would fall within this grade. It would also be likely that there would be other, possibly commercial, reasons for accepting a supplier in this grade. Perhaps, also, as does happen upon occasion, no other vendor is available.

VQA would be expected to work with those vendors in the middle grade to try to bring them up to the top grade. To the extent that vendors in the lowest grade have to be used, it would also be expected that VQA would work with them to help improve their performance so that they can be brought up to at least the middle grade.

It was said earlier that it is immaterial what kinds of "names" are given to the grades. Perhaps the most convenient would be either A, B, C or 1, 2, 3. Certainly, if computers are to be used, the single-character designations would be best.

Vendor rating began to be used about 30 years ago when published information was very scarce. What was available was limited to a very few schemes that required the use of computers and special programs and, in one case, a specially designed slide rule. Since those days there has been a very considerable advance in both appreciation and an understanding of the benefits that can accrue from the use of vendor rating.

The benefits that accrue when vendor rating systems are used can include considerable cost savings, to say nothing of intangible benefits, of which more will be said later. The early plans were concerned with rating a vendor on quality performance only, but it was not long before some users began to include other factors as well: factors such as price, delivery performance, general services, and in a few cases, product complexity. Most of these additional factors began to be included in work done in the United States. Of course, when other factors are included, it is necessary to make provision for suitably weighting them. In some instances in which these other factors were included, with unsuitable weighting, the outcome was that the quality factor accounted for less than half of the "value" of the actual rating determined. This would not be considered to be a satisfactory outcome. The quality factor *must* represent the major proportion of the rating.

For a variety of reasons, which should be obvious, it is strongly recommended that whenever consideration is given to the introduction of a vendor rating scheme, in the first instance at least, the scheme introduced should be as simple as possible. Only after experience has been gained in its operation and in making use of the results, and having gained knowledge of the problems that will undoubtedly arise, should consideration be given to the introduction of a more complex system which will include other factors. Potential users should be warned that this is a field in which it is well to "make haste slowly."

To provide a simple explanation of vendor rating systems and their method of operation, a system that I developed many years ago, when there was no published information about vendor rating, will be described. This system, which was remarkably effective then, is still as good and, perhaps more important, as easy to use as when it was devised. The benefits are also discussed.

When consideration was being given to the creation of this vendor rating system, a start was made by drawing up a set of simple but effective rules. It may also be mentioned that this plan was going to be used at the first European factory of an American company which is a world leader in its field. That set of rules, developed years ago, seems to be just as relevant today, and perhaps even more necessary, than when they were written.

1. The plan must be simple.
2. The plan must not require difficult calculations.

VENDOR RATING AND SURVEILLANCE

3. The plan must be suitable for use with the various standard sampling plans that are in use.
4. The plan must be able to take account of results from 100% inspections when they are used.
5. The plan must take account of the relative importance of faults.
6. The plan must show the difference in quality level of vendors and of different parts supplied by the same vendor.
7. The plan should be such that modifications can be made quickly and easily.
8. The plan should require the absolute minimum of labor to operate it.

An explanation of the reasons for these rules may be helpful. Rules 1 and 2 are fairly obvious, although with the generally increasing complexity of so many things, a plea for more emphasis on rule 1 would not be amiss. Rules 2 and 8 were necessary in this case because the only labor that was available to operate the plan was that of the quality manager—not an unusual industrial situation.

Rule 3 was necessary because in the particular circumstances in which the rules were written, some special sampling plans were in use as well as standard sampling plans. Rule 4 was also introduced because of these circumstances. Some of the parts concerned were of such a critical nature that 100% inspection was mandatory. Readers are advised that I was well aware of the criticisms leveled at 100% inspections. Nevertheless, in certain cases 100% inspections had to be carried out. The remaining rules are fairly straightforward and self-explanatory.

Rule 5 indicates that account must be taken of the relative importance of features to be inspected in incoming parts. To accommodate this requirement, four classifications were introduced. Three of them, from minor to critical, are the same as those given in Chapter 6. The fourth, and lowest, one was:

4. *Incidental:* A defect that will not affect product functioning in any way and is not likely to cause customer annoyance

With the passing of time it was felt that class 4 could be omitted; it therefore does not feature in further discussion of the system.

Although classes of faults had been determined, a little more had to be done to be sure that their relative importance was fully taken into account. This relativity was assured by the adoption of a weighting factor which was applied in the following way:

Critical faults: weighting factor 10
Major faults: weighting factor 4
Minor faults: weighting factor 2

(Originally, incidental faults were the base factor, with a weighting factor of 1.)

It must be stressed that those values were quite arbitrarily chosen in consultation with the company's engineering department and were considered to be appropriate at that time. In any other circumstances, any other values might be chosen and used. For example, in another, different scheme which I devised not long ago to determine levels of customer perception of quality, similar fault classifications were assigned weighting factors of 50, 25, and 1, respectively. In the different circumstances, and with rather different definitions, these weighting factors were considered satisfactory for their purpose. In any practical case the actual definitions chosen and the weighting factors assigned to them lie entirely within the discretion of the user. It is probably fair to point out that, upon later reflection, a more heavily biased weighting factor for the "critical" class would now be chosen if that early scheme was now to be operated. But it must be emphasized that the system *did* work and did provide the results hoped for.

The next step was the calculations. Everything is strictly arithmetical and simple, and no aids of any type are required, although now the ubiquitous pocket calculators will almost certainly be used. As the inspection of each shipment is completed, the number of faults that were found in each class were totaled and entered on the vendor part record card. The weighting factors were applied and the score for each class of fault was also entered on the record card. It is important to note that this vendor rating plan is based on the number of actual faults found, not the number of defective items in which the faults appeared.

So for each shipment there would be entered into the record card, among other items of information of the usual nature, two essential figures: first, the actual number of items inspected—this should correspond to the sample size, or the total number in the shipment when 100% inspection was applied; and second, the

VENDOR RATING AND SURVEILLANCE

total number of demerits (or weighted score) for all the faults found.

There was one record card for each part. Where one part was being supplied by more than one vendor, there was a card for each. Of course, the vendor's name appeared on the card. As shipments were inspected and the results entered into the cards, a bank of data was built up from the arithmetical summations made after every inspection was completed. The rate of accrual of data was such that it was decided that vendor ratings would be determined every three months. Of course, depending on the volume and the rate at which data accrues, shorter or longer intervals can be used. It will depend entirely upon individual circumstances.

When the time arrived for actual determination of the vendor rating, the procedure was also very simple. Three simple steps, only, were required, working from the part record cards:

1. Add up the total number of items that had actually been inspected.
2. Add up the total demerit score that those items had earned in total. (These two should already be on the cards from the summation done after every entry.)
3. Divide the total demerit score obtained according to step 2 by the number of hundreds of items actually inspected. (Step 3 was carried out solely to obtain numerical simplicity in the answer.)

The answer obtained in the simple calculation in step 3, in demerits per hundred items, is the actual vendor rating. However, a further simplification is made. The rating is a simple number which by itself does not really mean anything unless there is a scale against which it can be compared. Such a scale was produced and it is as follows:

100 or less: Excellent.
101 to 300: Good.
301 to 600: This vendor needs close attention from the quality department, as the performance is not good enough.
601 or more: This vendor is unacceptable; the purchasing department should be consulted with a view to finding an alternative vendor.

An example of how this plan works is as follows. In a period of three months the total number of items of one part that were actually inspected in the samples taken from the successive shipments amounted to 875. The total demerit score earned by those 875 items was 765. 765 divided by 8.75 (the number of hundreds of items) is 87. This is the rating of that part from that vendor. Therefore, the rating puts that part in the category "excellent."

This vendor rating system was applied to some 1600 parts and assemblies supplied by more than 200 vendors. It proved to a number of colleagues in other departments, especially the purchasing department, the importance and value of having such a system in operation, on the very day that the first rating figures were determined. One of its great virtues is that the "answers" it provides are based on fact, *not* on opinion. It is objective, not subjective. The example that proved the worth of the system on its first day of production of results is described next because it makes such an excellent case for the use of vendor rating systems.

A particularly dreadful "brick" had been dropped by one vendor. Sand had been found in oil pumps for diesel engines—a devastating fault. The immediate reaction among manufacturing, engineering, and purchasing personnel was that this vendor should be "struck off" the books without delay. However, as that was the day the first vendor rating results were going to be available, I was able to persuade my colleagues to hold fire until I had prepared a complete set of rating figures for all the items that this vendor was supplying—some 50 or more in all. By the next day, when all the ratings for that vendor had been worked out, a very different picture was apparent. All the parts that vendor supplied, except one, the oil pumps, were in the excellent category. Not only were they in that category, but they were about as good as they could possibly be. Only in the case of that one item were they otherwise.

Only the order for the oil pump was cancelled. The fuss immediately subsided as it was realized that the vendor was not nearly as bad as had been thought. They were actually very good and when the full story of the reason for the fault having occurred became known, one realized that it was one of those examples of human fallibility that happen even in the best organizations. But this happening demonstrated the value of vendor rating systems better than any recommendations could have done. They work on

VENDOR RATING AND SURVEILLANCE 125

the basis of hard facts and not on opinions, which can be very wrong.

As just discussed, vendor quality rating could be described as a passive form of vendor surveillance—surveillance operated from a distance in the bought-out quality audit area. Active surveillance is rather different and may be carried out in more than one way.

Perhaps the simplest method, and one that is certainly easy to carry out, is as follows.

Essentially, this form of surveillance is operated by means of a program of arranged visits to vendors. As discussed in Chapter 8, the CAST concept required close liaison between a customer and his or her vendors. In extending this liaison to what might be called the surveillance mode, the quality surveyor would arrange a program of regular visits to vendors. The frequency of visits to any vendor would depend on a number of factors, including the following: the volume of business, the "quality goodness" of the vendor, and the geographical remoteness of the vendor. But it should be noted that those vendors whose "quality goodness" is at the highest level should not be excluded from the program. They should not just be taken for granted.

These visits can be used for a number of purposes: to discuss incipient problems that the vendor may feel are future possibilities and to discuss possible future orders that the customer will be placing which might be of interest to the vendor from a materiel forward procurement point of view. This, too, is part of the CAST concept: to consider incipient problems that the customer feels might arise with the vendor's product. There are also other purposes, one of which is simply that of maintaining good relations, to show the vendor that they have not been forgotten and are still regarded as one of the "team."

There are some who say that such a program of visits to vendors is a waste of time and serves no useful purpose. In fact, the benefits that will accrue from such surveillance visits are of the intangible kind that improve long-term relationships, much as resulted from the program described in Case Study 3 in Chapter 8.

There is also a case for what might be called "reverse surveillance." This was a concept that I practiced during a period when I was working in West Germany as the European quality manager of a large international company supplying electronic components of some complexity to many customers in most European companies. I felt that there were many good reasons for maintaining good relations with my opposite numbers in a group of the most important customers in several countries. As a result, a program of regular visits to those customers was arranged. It proved to be very useful and resulted in "nipping in the bud" a number of problems which, at that stage, were minor. However, if they had not been taken up during a visit they could well have mushroomed to major problems. The personal relationships that were established resulted in an increasing readiness for the customers to give advance notice of incipient problems on an informal basis.

8
CAST
(Customer and Suppliers Together)

Over a period of many years as a quality manager in a variety of industries, mostly in large companies with worldwide reputations, I have seen many cases of what can only be called "adversarial situations" which have arisen between a company and its vendors. In the main these situations became apparent as a result of quality problems, although on investigation of those problems, it was sometimes the case that varying degrees of arrogance and/or ill will were shown to the vendor by a buyer in one's own company. Insofar as the vendors were concerned, because their main line of contact with a customer is usually through the buyer, the buyers appear to be the cause of their troubles. This is unfortunate for all concerned. It was also not uncommon for the attitude of the customer company's buying personnel to seem to be that of "big brother knows best" when dealing with difficulties raised by a vendor.

I have never been able to determine with any degree of satisfaction a single reason that could account for these adversarial situations. Colleagues, often in the buying department, who were apparently unwittingly creating the situations, appeared in all other respects to be normal, reasonable people who one was happy to work with. Yet adversarial situations were often created as a result of their involvement. In so doing they raised a "can of worms" that had to be dealt with by other colleagues who also had direct contact with the vendors.

Adversarial situations may arise for a variety of reasons. One important reason is that of a clash of personalities, for example a chief purchasing agent (CPA) who for no apparent reason sometimes displayed a very aggressive personality when dealing with vendors.

Case Study 1

A critical part was to be ordered and the CPA visited the prospective vendor to be sure that the vendor fully understood the critical nature of the part and that they were expected to supply it strictly in accordance with the specifications (which the vendor should do in any case). In due course the first shipment arrived and was inspected thoroughly. Everything was found to be correct except for one important feature. The part was required to be crack tested and free of cracks. From the number of cracks that were found at the incoming crack test, it seemed fairly clear that either the vendor had used unsuitable material or forgotten to crack test altogether. There could, of course, be many reasons for this failure, but the CPA took the bit between his teeth and, almost literally, stormed off to visit the vendor, announcing that "heads would roll." I never actually discovered if any heads "did roll," but on the return of the CPA there was no further mention of that episode and the vendor was given another chance. He performed satisfactorily thereafter.

In other cases the situation would arise when the customer company, for one reason or another, considered itself to be on a higher business level than the vendor. This may be thought to be the case because of company size or because of technology. It is also possible that the situation may be created by a vendor who believes he is in a monopoly situation with regard to his products. In this case the customer may be faced with a "take it or leave it" attitude. However, this "reverse" kind of adversarial situation does not arise very often and will not be discussed further, although it does no harm to recognize that it does exist.

It is usually the case that quality problems arise which have to be resolved if for no other reason than to ensure that the customer gets what he ordered. Apart from that, in the interests of good relations and the avoidance of further problems, it is not just desirable but, essential, to eliminate adversarial situations that are revealed. This entails the removal of the perceived causes to secure restoration of, or in some cases the institution of, the atmosphere of mutual trust, appreciation, and general cooperation that is so necessary for joint business success.

At this point, one should consider just what the consequences of adversarial situations are likely to be. The first and most obvious one is that of the general inconvenience caused by the rejection of incoming parts or even the discovery that defective parts have been assembled into one's own product. Delays in production schedules will result, with the possible further complication of having to remove defective parts from the end product. When replacement parts do arrive, it may well be the case that further delays will arise because of additional inspections called for because of the doubts raised about the ability of the vendor to provide conforming parts; and, of course, the feelings of frustration and ill will that will inevitably arise. All of these are inimical to the generation of good relations and the supply of conforming product.

Ideally, the relationship between a company and its vendors should be one in which both parties are working together consciously toward a common goal: that of customer satisfaction and joint business success. It is of no value to any company if its vendors deliver faulty product or even go bankrupt. To a vendor the converse is equally true. Therefore, to secure those aims it is necessary that there should be a good rapport between them: mutual trust, support, and an appreciation of the fact that it is in the best interest of both to avoid the creation of an adversarial situation. Such a situation automatically means problems; problems have to be solved and problem solving costs money—money that should not have to be spent.

The concept of CAST (customer and suppliers together) I conceived a number of years ago during the middle years of my service in various quality manager posts. The concept arose from a gradual intensification of my feelings after involvement in a number of adversarial situations within a short period. I had to soothe the feelings of several outraged vendors and generally "clean up the mess." The task was not an easy one, for many and obvious reasons, but it had to be done in the interests of intercompany harmony.

A detailed examination of the situations showed that the two factors of communications and technical information were major causes, apart from others mentioned earlier. The general quality of the technical information that was provided to vendors, and its general communication, was both poor and inefficient. The first caused uncertainty and frustration. Human nature being what it is, this combination of uncertainty and frustration resulted in the vendor turning in what, to the customer, seemed to be a poor performance. Thus the customer's appreciation of the situation was exacerbated and his general impressions of the vendor worsened.

It will be seen that at the heart of the concept, there is an understanding of the need for high-quality technical information and good communication—plus, of course, a realization on the part of both customer and vendor that they must work toghether in a partnership or team to achieve a common aim: customer satisfaction and joint business success. So, improvements have to be arranged in the system of communication and the quality of the information that travels along the communication system to the vendor. It was concluded that if these could be improved significantly, a major step in the right direction would have been taken.

Although the emphasis is on "teamwork" it should be said that the team can consist of varying numbers of members. It may range from a single representative each from a customer and a vendor; a customer representative and one from each of a number of vendors; or a number of representatives, each from a customer and some of his vendors. The actual composition will depend entirely on the nature of the particular problem and its significance. In an example discussed later, several representatives, each from the customer and a number of vendors, took part.

The case studies that follow deal with different aspects of the same overall problem, particularly communications. Each vignette pictures an adversarial situation, although there are a few points that must first be made. First, consideration should be given to the overall problem, and the actions that are planned should be as comprehensive as possible. Second, the reversal of the old to a new and acceptable situation should not be expected to take place overnight, or anything like it. Third, although outcome of the next two examples was considered to be satisfactory and moving in the right direction, it must be remembered that the corrective actions applied were devised for those particular situations. It does not necessarily follow that they could be applied directly to different situations, but they do provide guidance for others to plan suitable corrective actions to suit their own particular situation. Incidentally, even if there does not appear to be an adversarial situation, the CAST concept is well worth introducing.

Case Study 2

The large company involved in this case was involved in light engineering work. It employed several thousand employees, who produced a consumer product in high demand. Many vendors were involved and most were small companies, which tend to be more susceptible to communication problems. There were drawings and detail specifications for all the bought-out parts, although the requirements were not as demanding as those in Case Study 1 or 3. Nevertheless, many problems arose at incoming inspection and, not infrequently, on the product assembly lines themselves. And that was expensive. Many of the problems arose because of misunderstandings on the part of the vendors about the specifications and also because their lines of communication to the customer, through the buyers, often failed, partly, at least, because of a lack of technical knowledge on the part of the buyers. (It must be remembered that buyers cannot be expected to be technical experts in all technologies.)

The CAST concept was introduced by means of a major reorganization of the customer's quality assurance department. Within it, a new vendor quality assurance group was set up and staffed by a small number of engineers [vendor quality surveyors (VQSs)] who were given specialized training (the formation and operation of such a group is dealt with in Chapter 9). Each VQS was allocated a group of vendors on a geographical basis, including overseas vendors (of whom there were a number), for whom he assumed a "quality" responsibility. He received a copy of each order sent to one of "his" vendors so that he was fully aware of the requirements placed on them. Primarily, he acted as a "problem preventer" by maintaining a close liaison with each member of his group of vendors. Partly by telephone and letter and partly by regular visits. This liaison had the prime aim of ensuring, to the maximum possible extent, that impending problems did not become actual problems at receiving inspection, or, much worse, on the assembly lines.

The program was very successful and resulted in a considerable reduction in the number and severity of the problems experienced on the assembly lines caused by bought-in parts. The vendors were appreciative of the new facility that was now available to them. They felt that they were beginning to be part of the "team": CAST in action.

So far, so good. But it became apparent that there was another problem, which concerned the specifications that were sent to the vendors with the orders. Oridnarily, these were not passed to the vendor quality surveyors in advance. They relied upon their own departmental specification library for information about detail. For reasons that are not relevant here, the company produced large numbers of modifications to its specifications and drawings, and these did not always get to the vendors as they should. Dealing with, and resolving, those problems was beginning to take up too much of the time of the vendor quality surveyors. After further investigation it was concluded that a solution was possible with the willing and active cooperation of the buying department. Accordingly, a procedure was introduced which worked as follows.

The buyer prepared an "order pack" for each order to be sent to a vendor. It contained a copy of all the relevant drawings and specifications and any other technical information which would ordinarily be sent to the vendor with the order. Now, however, it was passed, in the first instance, to the relevant vendor quality surveyor, who had to satisfy himself on the following points before clearing it:

1. That all the drawings and specifications in the order pack were of the latest, correct issues
2. That all the drawings and specifications that were necessary for the vendor to complete the order were there
3. That no unnecessary drawings and specifications were there
4. That in all cases in which the principal drawings and/or specifications called up subsidiary drawings and/or specifications, their relevance was clearly indicated and referenced on and to the principal documents, and, of course, that they were in the order pack
5. That he was satisfied that with the drawings, specifications, and other documents in the order pack, the vendor had all the necessary technical information, and no more than that, to enable him to produce the parts as required

CAST

When the vendor quality surveyor was satisfied that all was correct, he initialed the cover note on the order pack and returned it to the buyer for posting to the vendor. This relatively simple action eliminated another awkward problem and reduced still further the number of problems that had been arising. There was further improvement in the general relationship with the vendors as soon as they appreciated that they were getting what might be called "certified data." Incidentally, the VQS was required to clear the order packs on a priority basis to ensure the absolute minimum of delay. Again CAST in action.

Case Study 3

This example is one that I like to think was imaginative and because of the nature of the problem, was also long term. The company, a world leader in its field, produced a complex, high-technology engineering product. Most of the component parts were brought in from more than 200 vendors for assembly and testing before shipping to the end customers. (The remainder were shipped in from the company's American parent factory.) Many of the parts were covered by very tight specification requirements. In fact, in some instances it was very difficult to find suitable vendors within the United Kingdom. Furthermore, strict conformance to the tight specification requirements was essential if the end product was to achieve the high standard of performance expected of it.

The average number of shipments received each month was about 1800, and a small number of the 200+ vendors could be regarded as specialist vendors. Nevertheless, from the company's point of view, the average percentage of rejected shipments was unacceptably high if one of the company's declared aims was to be achieved: to achieve a "no incoming inspection" regime for bought-out parts.

A careful study, including a quizzing of vendors, strongly suggested that apart from a lack of good communication, the vendors did not sufficiently appreciate the importance of the specification requirements (failure to appreciate the fact that specification requirements are intended to be met is an all too common failing). They did not know enough about the company, which was American and not long established in the United Kingdom. Nor did they have any idea of the procedures that were followed when their products were received and used.

To resolve the problem, I put forth a proposal which was considered in conjunction with a number of my colleagues: the chief purchasing agent, the chief engineer, and the manufacturing manager. It was agreed that a major vendor relations program should be established. After discussion it was also agreed that it would be put into operation by the writer and the chief purchasing agent, although most senior managers would participate in it. The program was to take place on a six-week cycle on the following basis.

In each case a group of four vendors would be selected, each of which made broadly similar products in a class that was giving trouble (there were plenty of these to select from). Each was to be invited to send a small team to the company's works for an intensive one-day briefing. They were told, of course, that the purpose of the program was the improvement of customer/vendor relationships. Each team was to include one each of the following types of personnel:

1. A member of the sales staff who had participated in the negotiations for the order
2. Someone who was directly involved in the manufacture of the parts
3. Someone who was directly involved in the inspection/quality control of the parts
4. In those instances in which there were specific metallurgical and/or chemical requirements, a member of the laboratory staff who had worked on those aspects of the parts

The program for the day was as follows;

1. Visitors to be collected from their hotels by company cars. (In most cases they had to take overnight journeys.)
2. Welcome from the managing director, with more detailed explanation of the overall program and its aims.
3. A detailed tour of the factory, from incoming material receiving through incoming inspection, stores, product assembly, to final test and preparation for shipping.
4. A film show. Two specially made company films were shown. The first demonstrated the heavy demands made on

the end product by users in particular applications. The second dealt with laboratory work and tests which showed the catastrophic consequences to individual parts if they did not conform to specification requirements.
5. Lunch at a nearby restaurant.
6. "Down to work." A session consisting of a series of short presentations by the following senior managers: chief purchasing agent, chief engineer, chief service engineer, and manufacturing manager. Each, in turn, discussed the consequences, from his individual departmental point of view, of component rejections or, much more important, of the incorporation of nonconforming parts into the end product.
7. Quality presentation. During this period the visitors were shown a series of examples of defective items recently received (ordinarily, there was no lack of examples) of the same class as those that they supplied, which sometimes were gross failures. Of course, all had been carefully selected so that none had been supplied by any of the companies present, nor were the actual makers identified of course. The nature of the problem was discussed in each case.
8. Open session. Here the visitors were invited to bring up for discussion any points that occurred to them about anything they had seen or been told, or any suggestions that they might like to put forward about improvements of any kind with regard to the parts they were supplying, perhaps by way of material substitution or cost reduction, or indeed, anything at all. This session was invariably welcomed by the visitors as a good round-off to a very valuable day. Nearly always at least one useful suggestion was put forth and later adopted.
9. Dinner at a local restaurant and return to their hotels.

After a few of these visits had taken place, it became obvious that a much friendlier atmosphere existed between the company and those vendors who had participated in the program. There was a much greater readiness to discuss problems before they became serious; there was more cooperation, more understanding, a greater willingness to discuss uncertainties before acting, and a real appreciation of the fact that both parties were actually seeking the same end—that, in fact, they were both part of the same team. The first major step toward the goal of the "no incoming inspection" regime had been taken: CAST in action.

Finally one should recall the example relating to shirts which was described in Chapter 3. That was an excellent example of CAST in action.

9
The Why and Wherefore of Vendor Quality Assurance

Before attempting to discuss the why and wherefore of vendor quality assurance, it would be well to define the term. It has been mentioned in earlier chapters and will appear again in later ones, although without going into much detail. Briefly, it is one technique for the control of quality of incoming supplies which may be used by buying organizations, more often the larger ones, for the following purposes: selecting, evaluating, and exercising surveillance over vendors, both active and prospective. In the performance of these tasks, it meets the requirements by means of the operation of a number of functions.

There are many reasons why the application of such a control technique should be necessary, in addition to the obvious one of not wanting to accept defective product from vendors. Another major reason is to reduce to zero, if at all possible, the problems caused to assembly lines when (accepted but) defective bought-out parts do not assemble. The lines may be brought to a complete halt. More serious still, the end product might be out on the market before it is discovered that defective parts have been built into it. Nor must one forget what is probably the most potent reason of all—the LOST COSTS—those costs which are usually unknown, and seldom recorded even if they are known, and which are caused by, for example, the cost of the search for replacement parts, the cost of halted assembly lines, and the cost of dealing with customer complaints about defective product. Examine the list of possible LOST COSTS given in Chapter 14.

The various functions sometimes operate independently of one another. But it is when they are operated together as a linked group that they form the control technique of vendor quality assurance (VQA). Its purpose, reduced to the simplest possible description, could be given as follows: consistent with the overall economics of a company, to ensure to the maximum possible extent that defective material is neither accepted by, nor delivered to, the purchasing company operating a VQA group.

When discussing setting up a VQA group, one should proceed in logical order. However, as it is very unlikely that any industrial company today would not have some form of incoming inspection already in operation, one does not need to consider the whole at this point. As incoming inspection under the new regime has already been considered in some detail in Chapter 6, only the other functions will now be discussed.

Not many years ago, the only regular method of making sure that a customer did not have defective bought-out parts from vendors built into his own product was simple inspection—goods inward inspection, in fact. The inspection methods were usually very simple and consisted of either or both of "percentage" and "100%" inspections. At best these methods could only be described as tolerable, and at worst as pretty poor. Because whatever they might or might not do to ensure that defective parts did not pass into production—and that was more than questionable—they did nothing to get to the root of the problem, that is, to stop defective parts from being made at all. Of course, later, there was a significant improvement in the efficiency of the overall inspection operation with the introduction of statistical sampling inspection.

However, it was not recognized, and is often still not recognized, that in the long run it is the customer company that pays for rejected products even if they are sent back to the vendor (and apparently not paid for). It is commonly said, although it is quite wrong, that the cost of rejected parts will be borne by the vendor. In fact, and in the long run, these costs will be taken into account by the vendor when presenting bills to customers. That is, the vendor will include the cost of rejected material in accounts rendered. This, of course, is never mentioned, but in fact, in the long run, it is always the customer who pays. The vendor will always adjust

WHY AND WHEREFORE OF VENDOR QUALITY ASSURANCE 139

prices to take account for all expenses, including the cost of rejected materials. Also, all the customers of a vendor will eventually pay for rejected material. The target for the customer should be: *No inspection is good inspection.* This aim is discussed more fully later in this chapter.

Over the years this fact of accounting life began to be accepted by the more progressive purchasing organizations, and a good deal of thought was brought to bear on the subject. It was realized that an ultimate aim must be to prevent defective parts from being produced at all. Some organizations began to take their first steps in this direction more than 30 years ago. But the biggest push, as far as the United Kingdom was concerned, came in the early 1970s. Then it became known that the Ministry of Defence (MOD) was proposing to make dramatic changes to the established goverment inspection systems—systems that had been operating with little basic change since 1919 when they were first established for the then fledgling Royal Air Force.

Taking account of the economic realizations that were coming to the fore (and the pressures for reductions in government expenditure), the MOD decided to adapt the requirements of the Allied Quality Assurance Procedures (AQAP). These procedures had been prepared for the use of the armed forces of the North Atlantic Tready Organization (NATO) countries and, in turn, were based on the U.S. government's military standards, of which the most relevant, and best known, is MIL-Q-9858, 1963 (See Appendix A).

The new system was introduced and came into force on April 1, 1973. Its principal effect was to place the complete responsibility for the supply of conforming product to the government squarely on the shoulders of the supplying main contractor company. Until the changeover, the government inspection agencies, of which there were no fewer than seven, had themselves accepted the responsibility and carried out the acceptance inspection. As far as the government was concerned, the change from the system that had been in use for 54 years was by means of the introduction of a system of evaluation and assessment of the contractor's quality organization. Furthermore, the basic rules that the contractor's quality organization would be required to follow left it with considerable latitude to adapt the rules as thought necessary to suit its own particular circumstances. In the first instance, the MOD would carry out an assessment of the contractor's organization using a team of assessors to determine if it conformed to the basic rules and was operating in the manner intended.

These new requirements were contained in a series of Defence Standards (Def. Stans.) in the group 05-21 to 05-29. Although they were applied only to companies in the defense industries, they attracted considerable interest in U.K. industry in general and also on a worldwide basis. As a result of this wide interest, the British Standards Institution (BSI) produced British Standard 5179, which was a general industry version of most of the MOD series 05-21 to 05-29 (MOD Standards 05-23, 05-27, and 05-28 were not included in the British standard). It is probable that these MOD Def. Stans. and BS 5179 were the foundation of most, if not all, of the vendor quality assurance systems being used by an ever-increasing number of countries around the world today.

More recently, BS 5179 has been replaced by the more-up-to-date BS 5750 (see Appendix A). This standard, in turn, is now being replaced by the new ISO standard series 9000 of the International Standards Organization (ISO). The ISO 9000 series will in due course probably replace a number, if not all, of other national standards in many countries around the world. These include those of Australia, Canada, the Federal Republic of Germany, South Africa, and the United States, to name a few of the better known ones. I strongly recommend that the first thing that should be done by any buyer who wishes to know about or become involved in any way in a VQA system is first to become familiar with the appropriate national standard, then later, when it is appropriate, with ISO 9000. BS 5750 is especially good for initial study as the world standard will be found to be very similar.

VQA as a whole carries out three primary tasks which operate in sequence: vendor assessment, vendor quality audit, and vendor rating. These tasks are described in some detail in Chapters 5, 6, and 7, respectively. Each has subsidiary tasks attached.

1. Vendor assessment. As will be seen from the detailed discussion later, there is more than one way in which vendor assessments can be carried out. Regardless of the method used, the outcome is a detailed examination of the capabilities and abilities of a vendor to meet customer requirements. Apart from quality, which is obvious, those capabilities may include other factors, such as relative price, ability to deliver on time, availability of the required product, and ability to meet special requirements should any be necessary.

There are two subsidiary tasks: first, the establishment of an initial grading for a vendor based on an assessment (this will be used as a guide when determining the individual tasks of vendor audit), and second, making arrangements for the approval of initial production samples prior to the commencement of bulk deliveries.

2. Vendor quality audits. This is a considerable development and extension of the old familiar "goods inwards inspection," which has been a feature of many company inspection/quality departments for many years. But today, vendor quality audits are much more than a routine inspection function. Operating on a planned and, according to circumstances, variable basis, they provide a continuous measure of the performance of a vendor and a confirmation, or otherwise, of the original assessment and grading.

Perhaps the only old, fixed feature about it is that there is still some direct inspection carried out. But these inspections are carried out as an audit function and their depth and scope will vary according to the degree of conformance to requirements of individual parts from individual vendors.

Here, also, there are two subsidiary tasks. The first concerns the arrangements which will ensure that suitable adjustments are made to the amount of inspection that is carried out. This will be according to the degree of conformance to requirements achieved by individual parts. If the degree of conformance is high, the amount of inspection called for will be reduced. If the degree of conformance is low, the amount of inspection called for will be increased. In extreme cases the inspection will be stopped, at least temporarily, and further incoming shipments of the part concerned will be refused. Until, that is, a solution has been reached to an obvious quality problem.

The second task concerns the record-keeping functions. This is an important part of the overall VQA control. These records are used for a number of purposes, of which the following are among the more important: confirming VQA assessments, providing evidence of a need for adjustments to the amount of inspection called for on individual parts, providing an overall picture of the entire VQA operation, and providing the basic data for the third VQA task, vendor rating.

3. Vendor quality rating. This technique has been known about and applied to an increasing extent over the past quarter of a century or more. As a matter of fact, I "cut my teeth" on the development and application of a simple and very effective system of vendor rating as far back as 1958. This is the system described in detail in Chapter 7. However, it is only in the past decade that vendor rating has come into its own with recognition of the fact that it is a valuable tool for evaluating the performance of vendors and exercising control over them. It is absolutely dependent on the records created from the operation of the vendor audit function and is, at least, a measure of the quality performance of a vendor.

This measure of the performance of a vendor can be compared with the original assessment, and if necessary, adjustments can be made to the method of assessment to produce more accurate forcasts. In much the same way that weather forecasters use actual weather results to improve the accuracy of their forecasting methods, the measure of performance, or rating, which is obtained may, indeed should, be used by buyers when consideration is being given to placing fresh orders.

In larger companies the scope of vendor ratings is often extended to take into account a number of additional factors not directly associated with quality. These would be regarded as being of special interest to buyers and were discussed in Chapter 7. In such a case, of course, the system name would change to vendor rating and it would present to buyers an overall picture of the "goodness" of a vendor.

It must be emphasized that one exceedingly important aspect of the quality content, at least of vendor rating, is that it is based entirely on factual data. When factual data are not available and only "smoke" is visible, too much attention is paid to the old saying: "Where there's smoke, there's fire." As a result, cases have been known in which a vendor has been "forever damned" as a result of a single error, as in the case of the oil pumps described in Chapter 7. In most cases, if factual data of the kind that vendor quality ratings provide are available, vendors would not be condemned unjustifiably and out of hand.

A decision having been made to create a VQA group, it would be obvious that among other things, staff will have to be recruited and

WHY AND WHEREFORE OF VENDOR QUALITY ASSURANCE   143

suitable office accommodation will have to be provided. "Suitable" does not just mean desks and chairs in some odd corner of an office. It must be remembered that one of the aims of a VQA group, albeit a rather minor one, is to create an impression in the minds of representatives of vendors that the group is well organized and knows what it is doing. This aim is achieved partially by the provision of good, if not actually luxurious office accommodations, in which the quality surveyors are provided with pleasant work surroundings. It may be that the periods they spend in the office are limited, but when they are in and having discussions with representatives of vendors, comfort and appearance are very desirable.

Good quality surveyors are likely to spend a large part of their time on the road, but much of their "in" time will be spent with representatives of vendors. This clearly points to a conference room as a necessary part of the suite of offices, and, of course, a separate office for the VQA manager. As the obvious is often overlooked, it may be well to mention that generous provisions should be made for the filing capacity that will be required. In the course of time there will be a considerable accumulation of paperwork, to which frequent reference is likely to be made. Of course, the VQA suite of offices should be as near the existing BOQA or incoming inspection area as is practicable.

Measuring the effectiveness of the operation of a VQA group is an important, although difficult, factor to handle. Perhaps the simplest, and certainly a very obvious method would be on the basis of the number of problems or actual stoppages experienced on the assembly lines or equivalent manufacturing area. In parallel with the program for setting up the VQA group, arrangements should be made for the gathering together of data, mostly historical and with the assistance of the existing quality organization, about the problems that have been experienced on the assembly lines and in press and machine shops, that is, problems caused by nonconforming bought out parts or materials. Whatever the magnitude of these problems may have been, they will form the baseline from which the effectiveness of the VQA operation can and will be judged.

Areas of operation will have to be determined and allocated to individual quality surveyors. Of course, there are many ways in which this can be done, and the home locations of the surveyors should be taken into account if at all possible. (It will often be the case that the surveyors will operate from their homes.) This may substantially reduce the amount of traveling time that has to be undertaken, thus reducing costs. However, these costs may be considered to be relatively minor. When finally deciding upon work allocations, there are three factors in particular that should be considered:

1. The expertise of individual surveyors
2. The variety of types of items being bought and their relative complexity
3. The geographical locations and spread of vendors

The allocations that are finally made will depend to a large extent on the particular circumstances and the preferences of the VQA manager. However, as cost is always an important factor to be considered, it is suggested that geographical location should be the basis, to some extent at least, of the plan adopted. This would be of particular importance when there are foreign (overseas) vendors. It would not make sense if a rigid work allocation by product types resulted in two surveyors having to pay simultaneous visits to two vendors in the same town, especially if that town happened to be in another country. But it must be emphasized that all the related factors should be taken into consideration even if the final choice has to take into account local circumstances and preferences. The best choice in each case will be that which best suits the circumstances.

It is desirable to reemphasize that a most important factor in the operation of a VQA group is that of costs: both costs incurred, which are important and easy to determine, and costs saved, which are even more important and much more difficult to determine. By the very nature of its operation the direct VQA costs will be significant, partly because of the need to provide company cars for the surveyors but also because of the increasing need for foreign travel. In the particular VQA operation that has been the model for most of the descriptive examples in these chapters, vendors were located in almost a dozen European countries on both sides of the "iron curtain" as well as some in the Far East. Not surprisingly, perhaps, the latter did not normally receive visits more frequently than once per year!

On the credit side one group of costs is relatively easy to establish. This is the cost of the average number of labor hours lost because of bought-in parts and material found to be defective on

WHY AND WHEREFORE OF VENDOR QUALITY ASSURANCE   145

the assembly lines. In establishing those costs it is very important to remember to include the overhead charges on those labor hours. They will usually be quite a lot higher than the direct costs. Past experience strongly suggests that these overhead charges will often be ignored or "forgotten." In one example that I investigated, a direct cost of £35 was charged as a loss due to defective parts. Actually, when the total of all the related indirect and overhead charges was taken into account, the cost charged in this particular case should have been almost £300! *Never forget the overhead charges*. However, those direct costs will usually be determined on the basis of labor hours lost.

After the VQA operations have begun and have had an opportunity to settle down, it should be possible to make a statement such as the following:

> The VQA operation has been able to reduce the average number of labor hours lost on assembly and other operations which were caused by nonconforming bought-in parts during the last month by X%. This is equal to $. . . . "

Don't forget to express all savings in money terms, and don't forget the overhead element. Even if this particular measure is not used, it is important that some measure is determined and used. This is because it may be taken for sure that the VQA group will be expected to justify its cost later, if not sooner. Experience suggests that it will usually be sooner. Other costs savings can be shown to be due to lower costs of disruption arising from fewer rejected incoming shipments at BOQA.

Before leaving the question of costs for the time being, attention can usefully be drawn again to two sayings that used to be well known in the field of quality:

1. All inspection costs money.
2. No inspection is good inspection.

One of the basic aims, perhaps the most fundamental, of a vendor quality assurance group should be to recognize the validity of the first of those sayings, and work toward the achievement of the second, at least as far as bought-in parts and materials are concerned. In other words, a long-term aim of VQA should be to reduce the need for incoming inspection to zero. In theory, at least, this goal could be reached by a combination of good selection of vendors and assistance in helping them to improve their systems for their quality control. This assistance could, of course, come by application of the CAST concept discussed in Chapter 8.

In concluding this chapter it should be emphasized that the general details of operation of a VQA group which have been described were taken from the real VQA group in the large light engineering consumer product company discussed earlier. The group was very successful.

Postscript. It is rather unusual to find a postscript at the end of a chapter in a book. In this case I feel that it is warranted because of a comment I read on the very day that the draft of this chapter was completed. The essence of this comment is as follows:

> Experience in the production of high-reliability subassemblies strongly suggests that, even when PPM (parts per million) schemes are in operation by a supplier, one will not be able to dispense with the operation of an intensive BOQA function. This is because, it is suggested, the costs of correcting faults in service due to faulty components that may have gotten through PPM procedures and normal audit procedures at the customer's end is becoming astronomically high.

A short note about the parts per million philosophy may be in order here. It was developed in Japan and has been used increasingly the past five years or so. It is probably most closely associated with another comparatively recently introduced practice, that of statistical process control.

It is claimed that many Japanese companies are supplying components in large volume, with the proportion of rejects no more than that "parts per million." The lowest figure that I have heard about is 10 or fewer defectives per million, although 100 to 500 is more common. In terms of "old-fashioned" acceptable quality levels (AQLs), that would be about 0.05%. Looked at in this way, it may not be quite as much of a "dream" after all. However, these results are obtained with the aid of that statistical process control. I consider that statistical process control is no more than an updated version of the statistical process control that has been used all over the world since the late 1920s.

I believe that these techniques are more likely to apply to companies that manufacture their product in very large volume. However, I also believe that, provided that their manufacturing processes have ample capability to maintain required tolerances and they can control them closely, smaller companies should be able to approach that Japanese "dream."

10
The Importance of Specifications to BOQA and VQA

There are two classes of documents that are vital for the effective operation of a VQA system, specifications and procedures. However, there is often a degree of confusion about the meanings that are given to each of these terms. So, for the purposes of this chapter, and indeed for the whole book, the following definitions are the ones assumed throughout.*

Specification. A specification is a document that contains a detailed description of the particulars of a required item, including dimensions, materials, functional requirements, and so on. It will also contain any other information that will be required by whoever is going to supply the required item to ensure conformance to the requirements. It may be written, or a drawing, metallurgical, chemical, or environmental.

Procedure. A procedure is "the manner of proceeding or going on"; a detailed description of a function and the way in which it should be carried out, together with any other relevant information.

It may be said that without effective specifications and operational procedures, a VQA system will be useless. Without effective specifications it will not be possible for a vendor to be sure what it

---
*In part, these definitions come from the *Oxford English Dictionary*.

is that he is to supply, nor will it be possible for a customer to determine satisfactorily if a vendor has conformed to requirements. Without suitable and effective operational procedures, the VQA personnel cannot be sure that what they are doing is what they ought to be doing, or if they are doing it in the required manner. (Management systems in general could not operate without specifications and procedures, either.)

There is a good deal of evidence to the effect that the expanding production of (unclear and ineffective) specifications within industry is causing more and more problems, and, of course, more and more examples of poor specifications are coming to light. Although the definition of a specification given in this chapter is clear, there is an amendment that should, perhaps, be made. The "detailed description" may be either pictorial, as in an engineering drawing, or written in words, as in an environmental requirements specification, or any combination of the two.

It is unfortunate that the problems that arise can be found in both pictorial and written versions (although more commonly in the written ones) and that there is no easy solution to the difficulties. In the case of pictorial representations, and even here one can find many obvious departures that can confuse the vendors, conformance to the recommendations contained in BS 308 "Engineering Drawing Office Practice" (see Appendix A) or equivalent, should be the aim.

Many problems are caused by related information included in a drawing in the form of written descriptions of, for example, treatment of finishing processes, although one will often find dimensions on a drawing that are irrelevant to the functioning of the item. They are, in fact, necessary for tooling purposes, but no indication of this fact is given, with the not uncommon result that a vendor will strive very hard to meet a dimensional requirement that is really of no consequence to the vendor or, in fact, to the finished part itself. In addition to the problems of conformance that may arise, there can be little doubt that unwanted costs arise as a consequence.

Whether the specifications will be mainly pictorial or mainly written will depend largely on the industry. In the case of engineering industries, the drawing type is likely to be predominant. In electronic and high-technology industries in general, the written type of specification predominates and can be expected to be quite

THE IMPORTANCE OF SPECIFICATIONS TO BOQA AND VQA 149

troublesome. Some of the difficulties that arise do so because the specifications writer is "too clever by half." The writer may be trying to impress the reader/user with the high standard of his or her knowledge but, in fact, may only succeed in confusing, irritating, and annoying. One example of this kind is given by the following definition of "acceptance number":

> *Acceptance.* The acceptance number is defined as an integral number associated with the selected sample size which determines the maximum number of defectives permitted for that sample size.

(Actually, it does not "determine" the maximum number of defectives permitted; it *is* the maximum number.) Most people reading that definition would be likely to agree that there must be a simpler definition.

The operational procedures that describe the work of the VQA group may form part of the company's quality manual. But they may also, instead, form part of a separate and specialized vendor quality assurance manual. There will be a number of procedures and they will cover all the aspects of the work of the VQA function. Typical examples of VQA operational procedures will be found in the appendixes to this chapter. Some of the titles would be as follows:

Selection, Assessment, and Grading of Vendors
Approval of Initial Production Samples
Surveillance of Approved Vendors
The Bought-Out Quality Audit
VQA Record Keeping
Sampling Plans for the Bought-Out Quality Audit
Quarantine and Bonded Stores

Other titles will come to mind, but these are representative of what one would expect to find covering the activities of a VQA group.

There are a number of ways in which the preparation of operational procedures for a VQA group may be undertaken. In general, it does not matter very much which method is used except for one important point. There must be a standard format of presentation and a uniform writing style throughout. Uniformity of format is relatively easy to achieve, although uniformity of writing sytle is not as easy. There are two basic ways by means of which this objective can be achieved, and the best is to have a single person write them fully from draft to final form. If this is not possible for any reason, an alternative is to accept drafts from any source and have the final versions written by one person. There is little doubt that a series of documents of a related nature which have to be used by technical and shop floor personnel will be much easier to read and understand if there is a uniform format and writing style. If not, problems will not be long in arising.

In many company organizations there is no formal scrutiny of documentation packages to vendors to ensure that:

1. The right documents are sent
2. The documents are not ambiguous and confusing
3. All of the required documents are included in the package.

It has already been said that it is far from unknown for the wrong documents to be sent, or for those sent—or some of them—to be ambiguous and confusing. But another problem is that, quite often, not all the required documents are sent to the vendor. The vendor has to search through all that he has been sent to find out exactly what it is that he is supposed to do, and then sometimes, he finds that he is referred to another essential document which he does not have and which has not been sent to him.

One example of this kind that I experienced concerned an inquiry sent out by a very large buying organization. The details included a long and very comprehensive specification. A kind of "ring around the mulberry bush" situation arose. This was because, after reading this very long specification, the prospective vendor discovered that he had to ask for another one, which appeared to be very relevant but had not been sent. This supplied, reading of it revealed the need for a further three specifications. These, too, were requested and supplied. Upon reading them it was discovered, to the surprise of the vendor, that the first one referred back to one of the earlier ones, the second one was not relevant after all, and the third referred to further specifications. This is far enough to go in this description, but there were more steps in the chain. The entire situation was very confusing indeed and not calculated to induce the best response from any vendor, however well intentioned he might be toward the prospective customer—to say nothing of the LOST COSTS.

Although this was an extreme case, this type of situation seems

THE IMPORTANCE OF SPECIFICATIONS TO BOQA AND VQA   151

to arise with almost monotonous regularity just because little, if anything, is done to prevent it from happening. The question that immediately arises, of course, is whether there is anything reasonable that can be done. Of course, the answer is that there is. The procedure described in Chapter 8 as the second part of Case Study 2 is a most effective answer.

One other type of specification that must be discussed here is the audit inspection specification. When audit inspections are carried out on incoming parts, it is most important that the inspections be planned. Indeed, this should be the case for any inspection. It should be no part of the task of an inspector to decide what to inspect nor how to do it. The task of the audit inspector is to determine if the part conforms to the requirements of the specification. It is not, and never should be, the task of an inspector to determine if nonconforming parts can be used. Either the decision is an automatic one depending on the facts produced by the inspector, or the decision will result from a referral to a qualified person when nonconformance is found.

It is therefore essential that inspectors have precise instructions and directions which tell them exactly what they have to do in the various circumstances. In general, this will be done through the medium of the operational procedures. In the case of individual parts, however, the vehicle for instructions for audit inspection will be the part inspection specification. This may also be called an acceptance specification.

There are many types of such specifications, but there is one common factor. Each must contain all the necessary information to tell the audit inspector exactly what it is that he or she has to do. In part, it may consist of a dimensioned and toleranced drawing, but it should also contain other information of the following types:

1. Identification of the various features that have to be inspected, and their relative importance from critical to minor
2. Identification of any gauges or similar devices required, and for which features they are to be used
3. Instructions about any sampling plan to be used
4. Disposal instructions for a shipment according to the results of an inspection

5. Standard information about the recording of the results of all inspections

Depending on the individual organizational circumstances, the contents of the audit specification may vary, but the content list just given can be taken as typical of what would normally be required.

The reader should by now be in no doubt about the importance of specifications for the effective operation of a VQA group. Although in a sense this requirement parallels that in Chapter 2, it is of much greater importance here. Appendix 1, from Company A, presents a procedure describing the government requirements for quarantine and bonded storage in which BOQA and the VQA surveyors would have a close interest. Appendix 2, from Company C, is a procedure that describes the approval of supplier initial production samples.

APPENDIX 1: PROCEDURE FOR QUARANTINE AND BONDED STORAGE

1. Purpose of Procedure. To describe the requirements of the British Ministry of Defence for quarantine and bonded storage and to detail the method by means of which they will be met.

2. Scope. The procedure concerns the production control sections of both manufacturing divisions of the Company, which have direct responsibility for operating the storage facilities, and also the QA department.

3. Related and Relevant Procedures and Other Documents

Def. Stan. 05-21, paragraphs 210c, 215 and 217b
04-601-001   Handling, Segregation, and Identification of Materiel
05-613-001   Concessions and Production Permits
06-101-001   Control of Nonconforming Materiel

4. Responsibilities. As the responsibility for the direct operation of storage facilities in the Company is vested in the production control sections of the manufacturing divisions, the prime responsibility for providing them rests with those sections. However, as

THE IMPORTANCE OF SPECIFICATIONS TO BOQA AND VQA

the QA department has a responsibility for ensuring that requirements of this procedure are met, there is a joint responsibility to this degree.

5. **Requirements of the Procedure**
   5.1. *Definitions*
      5.1.1. "Quarantine" storage is an area to which access is restricted and controlled by the QA department and which is used for the following purposes:
         5.1.1.1. Material that is received into the factory for direct productive purposes or for special production processes must be "quarantined" until such time as the incoming inspection to determine conformance to the requirements of the governing specification(s) has begun. When, for any reason whatsoever, only a sample of the material is taken for the incoming inspection, the bulk must remain in quarantine until the inspection has been completed satisfactorily, when it can be removed into the bonded storage.
         5.1.1.2. If material that is passing through the manufacturing cycle is inspected at any stage and some or all of it is found to deviate from the requirements of the governing specification(s) (i.e., it is nonconforming), the nonconforming material must be "quarantined" until such time as a decision is made concerning its disposition. If the nonconforming material is clearly scrap, it need not be quarantined, but it must immediately be disposed of as such.
         5.1.1.3. If, for any reason, material that has previously been inspected becomes suspect, it must be removed from its current location and placed in "quarantine" until such time as its integrity is confirmed or another decisioin is reached.
         5.1.1.4. Quarantine storage contains only materiel that has not been inspected or which, having been inspected, has had doubts raised concerning its acceptability and which is quarantined until its acceptability has been demonstrated. But it will also contain material that is known to be nonconforming which is being held there until a decision is reached about its disposition.
      5.1.2. "Bonded" storage is an area to which access is also restricted, as for quarantine storage, which is used for the following purposes:

5.1.2.1. Material that has completed a manufacturing cycle and has been inspected and found to be acceptable to the governing specification(s) must be placed into bonded storage until it is required for further production purposes.
5.1.2.2. Productive material that has been received into the factory, or other material, as in paragraph 5.1.1.1, and has been inspected and found to be in conformance with the governing specification(s) must be placed into bonded storage until it is required for further production purposes.
5.1.2.3. All finished parts that have been inspected and accepted must be kept in bonded storage until they are required for further production purposes or shipping.
5.1.2.4. Bonded storage may only contain material that has been inspected and found to be in conformance with the governing specification(s).

5.2. Subject to the general requirements concerning access, restriction of access, and general controls over access, the location, size, shape, or form of the quarantine and bonded storage areas can be to suit operational convenience and the nature of the material that is going to be in such storage.

5.2.1. Quarantine storage may be locked boxes or cupboards as well as secure and large fenced-off areas of a factory. Bonded storage may be a section of a larger general storage area and clearly delineated as a "bonded" area as well as being separate storage sites in their own right.

APPENDIX 2: PROCEDURE FOR SUPPLIER INITIAL PRODUCTION SAMPLE APPROVAL

1. Purpose of Procedure. To describe a system for the approval, by the supplier quality assurance (SQA) group of the Company, of initial production samples from suppliers.

2. Scope

2.1. This procedure applies to all orders for production material placed with a supplier for the first time and to all reorders where new tooling is brought into use.

2.2. At the discretion of the Company, samples for approval may be called for against any order for production material at any time.

THE IMPORTANCE OF SPECIFICATIONS TO BOQA AND VQA

2.3. The procedure is primarily of concern to the SQA group and the purchasing department. There is also involvement on the part of the design department, the laboratories, and the component measuring center (CMC).

3. Related and Relevant Procedures and Other Documents. These are to be determined later.

4. Responsibilities.

4.1. The following departments and groups of the Company have various responsibilities according to the requirements of this procedure.

Purchasing department
SQA group
Design departments
Laboratories
Component measurement center (CMC)
Concept design

5. Requirements of the procedure

5.1. *General*

5.1.1. For the convenience of users of this procedure, the main sections of relevance are as follows:

5.1.1.1. *Purchasing Department:* 5.2, 5.4.6.2, 5.4.6.2.1, 5.4.7.2, 5.4.7.2.1, 5.4.7.3, 5.4.7.3.1

5.1.1.2. *Supplier:* 5.2.2 and subsidiary paragraphs, 5.3

5.1.1.3. *SQA Group:* 5.4

5.1.1.4. *Design Departments, Laboratories, CMC:* 5.4, 5.4.4, 5.4.4.1

5.2. *Purchasing Department*

5.2.1. When the purchasing department first approaches a prospective supplier, they will ensure that the supplier is given a copy of the SQA manual, which contains, among other data, the following items of information:

5.2.2. A minimum of 10 initial production samples must be provided, 20 samples for transfers. If more are required, ordinarily, they will be specifically requested. However, if multiple molds, dies, or tools are used, there must be a minimum of 10 samples from each impression or copy, and each sample must be identified against the relevant impression or copy.

5.2.3. The "Initial Production Sample Report" (form CQ.012) must be completed by the supplier and sent with the samples.
5.2.4. Each individual sample must be identified with one of the labels provided.
5.2.5. When samples are ready and identified, they should be sent, with the completed form CQ.012, to the Company for the attention of the SQA group.
5.2.6. On no account must production quantities be sent against the order until the supplier has received written approval of the samples from the Company buyer and delivery schedules have been received.
5.2.7. If at any time the supplier wishes to ask the Company to consider the acceptance of material that does not meet Company specifications in full, the supplier should submit a request for consideration, with full details, on the "Supplier Request for Concession" form. The form may be obtained from, and should be sent to, the purchasing department.
5.2.8. If the supplier wishes at any time to request a change to a Company specification, the supplier should submit full details on the appropriate "Request for Change to Specification" form. This form, also, may be obtained from, and sent to, the purchasing department.
5.2.9. Before actually placing an order covered by the requirements of this procedure, the buyer must confirm the following:
   5.2.9.1. Design clearance for the item has been given.
   5.2.9.2. The proposed supplier has been approved by SQA.
5.2.10. In general, all contact between a supplier and the Company should be via the purchasing department. However, when technical queries arise, SQA will deal direct with the supplier but will keep the purchasing department advised.
5.2.11. When the order is sent, the buyer will ensure that copies of the form CQ.012 and sample identification labels and also sent.
   5.2.11.1. If the order falls into the category described in paragraph 2.2, the buyer will send a covering and explanatory letter with the order together with the necessary form CQ.012 and sample identification labels.
   5.2.11.2. Concurrently with the dispatch of the order, the

THE IMPORTANCE OF SPECIFICATIONS TO BOQA AND VQA

buyer will prepare an "advice of order" memo in duplicate and send a copy to SQA.

5.2.12. The buyer is responsible for progressing the order until the initial production samples have been received. Samples should in general, be received some 2 to 4 weeks before the planned delivery date for the first bulk delivery.

5.3. *The Supplier*

5.3.1. The supplier must fully comply with the requirements described in paragraphs 5.2.2 to 5.2.8.

5.3.2. In checking the initial production samples, the supplier must ensure that each is checked for conformance to the specification(s).

5.3.2.1. The supplier must also ensure that the highest and lowest measurement for each feature on the items is recorded on the form, CQ.012, on which all entries must be typed. Any form not completed correctly will be returned with the accompanying samples.

5.4. *Supplier Quality Assurance*

5.4.1. When a set of samples and the associated form is received from a supplier, SQA will complete and return the "advice of order" memo to the purchasing department to advise them that the samples have been received.

5.4.2. SQA will enter details of received samples in a log for that purpose.

5.4.3. SQA will divide the sample as they consider necessary and send those part samples, each accompanied by the appropriate copy of form CQ.012, to the relevant design department, the laboratory, the component measuring centre, and where necessary, to the concept design group.

5.4.3.1. SQA will be responsible for all progressing that may have to be done after receipt of the samples to ensure rapid processing.

5.4.4. The relevant design department, the laboratory, and the CMC will make the necessary examinations of the part sample items sent to them as quickly as possible. They will enter the results on their copy of form CQ.012, together with their overall conclusions about conformance.

5.4.4.1. When they have completed their examinations of their part samples, the design departments, the laboratory, and the CMC will return their part samples, if they have not been destroyed during testing, with the completed forms to SQA.

5.4.5. As the part samples and the completed forms CQ.012 are received back from the examining centers, SQA will record the individual decisions in their log (see paragraph 5.4.2).

5.4.5.1. When all results have been received, SQA will formally make, and record in the log, an "accept" or "reject" decision and then take the appropriate actions for either decision as described in the following paragraphs.

5.4.6. *Acceptances*

5.4.6.1. SQA will prepare three copies of the standard "acceptance of samples" letter. All three copies will be sent to the buyer concerned and SQA will file all three copies of the completed CQ.012 form. An approved sample will be sent to bought-out quality audit (BOQA) and to the supplier.

5.4.6.2. Upon receipt of the three copies of the "acceptance" letter, the buyer will note the fact in the records and distribute the three copies as follows:

Copy 1 to the supplier for his information
Copy 2 to the initiator of the order
Copy 3 to be filed by the buyer

5.4.7. *Rejections*

5.4.7.1. SQA will prepare three copies of the standard "rejection of samples" letter. They will also enter full details of the reason(s) for rejection as given on the relevant copy, or copies, of form CQ.012. All three copies will be sent to the relevant buyer and SQA will file all three copies of the completed form CQ012.

5.4.7.2. Upon receipt of the three copies of the "rejection" letter, the buyer will proceed as in paragraphs 5.4.6.2.

5.4.7.3. With the copy of the "rejection" letter that is sent to the supplier, the buyer will enclose a further set of forms (CQ.012) and labels so that the supplier may submit further samples for approval.

5.4.7.3.1. The buyer will also raise a new "advice of order" memo, as in paragraph 5.2.11.2.

5.5. Further samples will be dealt with as in paragraph 5.4 et seq.

11
Vendor Quality Assurance, Liaison, and Backup

Once again much will depend on the individual circumstances of the company, but it is most desirable that the VQA group be supported and backed by the full strength of the quality department. In a company that buys a high proportion of its basic production needs for incorporation into, or conversion to, its own end products, the VQA group will be very important from the point of view of the overall performance of the company. During the earlier stages of its operation, it is quite likely that the reason for its being, let alone its actual existence, will be questioned by existing "vested interests" in the company structure. It will be looked on by some as an additional and unnecessary overhead expense. Therefore, it will need all the support that it can get, at least during its formative period, to enable it to do its job. As time goes on and it proves its worth, this level of support will not be as necessary and the group will be able to stand on its own feet. It will have shown that it is earning its keep.

Of the other departments in the company with which it needs to maintain close links, perhaps the most important is the buying department. This is dealt with shortly. But for now consider the links with the rest of the quality department.

Quite apart from, perhaps, moral support, a VQA group needs considerable technical backup if it is going to do its job properly in the technical sense. In another sense it is acting as an executive arm of the design department, especially when prospective vendors are being assessed to determine their capabilities for the production of a new item. Quite apart from this aspect, there is considerable need for access to other, highly technical support: for example, metallurgical and chemical laboratory facilities and, if possible, a center for the examination of sample components, the last usually by direct measurement methods as opposed to gauging, and very intensively as compared to routine audit inspections. Upon occasion, also, there may be a need for guidance in statistical techniques or direct assistance from a statistician. There must also be a close relationship with the remainder of the quality department, as well as lines of communication with other departments of the company.

One can assume that any company which possesses a quality department and is now setting up within that department a VQA group will already have ensured that the quality department will have all the usual sections. These are likely to include the following, with all of which VQA will have close links:

Incoming inspection
Product quality assurance
Standards laboratory
Statistical section
Laboratory facilities (may be in another department)

The general nature of the relationship between the VQA group and the various sections of the quality department will vary but will probably be on the following lines. Incoming inspection should be completely integrated with VQA and, in fact, become bought-out quality audit (BOQA). There is such a degree of interrelationship between their activities that it would not make sense to keep them separate. Also, the amount of direct inspection that it will carry out will depend to an extent on the work of the VQA surveyors. Conversely, VQA will depend upon BOQA for advance signals about incipient problems. The work of the two is complementary.

A proportion of the problems that are experienced by the product quality assurance section, which has a responsibility for the outgoing product quality, will depend on the work of the VQA group and its effectiveness. Problems arising in assembly departments because of nonconforming bought-out parts are a frequent cause for complaint in industry. This emphasizes the need for

VENDOR QUALITY ASSURANCE, LIAISON, AND BACKUP

close links between product quality assurance and VQA. It would be expected that the former would keep VQA informed about the problems that it experiences because of bought-out parts. Conversely, it would also be expected that VQA would alert product QA whenever it appeared that there might be a problem about undetected nonconforming parts or materials so that action could be taken to prevent an assembly problem from arising.

All the laboratories, chemical and metallurgical on the one hand and standards on the other, provide highly specialized services which are very important to the VQA group. Nevertheless, it would be unreasonable to expect VQA to have such laboratories on its own. So it is to the chemical and metallurgical laboratories, perhaps under the control of a department other than quality assurance, that VQA will look for all its necessary information and guidance about the purely physical elements of bought-in parts and their various treatments. Similarly, it will be the standards laboratory that will carry out to a high degree of accuracy the dimensional checks on, for example, initial production samples, unless, of course, as was suggested earlier, there is a sample component measurement center. Assistance from the statistical section is likely to be required less frequently, although there will be "friendly relations."

In the foregoing discussion about the relationship of the VQA group with the rest of the quality departments, problems would not really be expected, or likely, as they are all under joint control. In the real case that is the basis for all these discussions, the overall control was in the hands of a director of quality and procurement. But one must now consider the relationship that VQA must have with the buying department when each is under separate control, and the role of individual buyers in this relationship.

It would not be surprising if, at least during the early stages of the work of a new VQA group, there were a degree of antipathy, at the least, on the part of the buying department as a whole, or individual members of it. To set up a new group that will have close links with vendors and work closely with the department, especially when it is under separate control, immediately suggests implied criticism of the previous part played by the buyers in selecting and dealing with vendors. This feeling is likely to arise, at least partly, because it has long been considered that it is the buyer alone who has responsibility for the selection of vendors.

It may not help very much in the first instance, either, to explain that the surveyors will be doing things that were never done before but which are intended to help buyers with the process of vendor selection. It might help even less to suggest that the new techniques and skills that VQA will employ are both necessary and ones that buyers do not normally possess. If, however, a persuasive person is given the task of explaining the entire VQA operation and how it is intended to work, there is every likelihood that the initial antipathy will evaporate, especially if the explanation "works backwards" from problems experienced on the assembly lines and explains that buyers are an essential part of the system.

An additional very helpful action that could be taken when setting up a VQA group would be to second a senior buyer to the quality department to help set it up and act as its manager until it has settled down and is running smoothly and an opportunity arises to recruit a technically qualified manager. This will provide two invaluable reassurances for the buying department: First, it demonstrates how much they are being relied upon, and second, how the VQA operation supplements and complements theirs. A further invaluable by-product of this arrangement is that the senior buyer concerned receives unique quality experience which should be of great benefit to the buyer, personally and, indirectly, to his or her buying colleagues. This was the tactic that was most successfully employed in the real case already referred to.

In the course of the normal operation of a VQA group, it must be reiterated that there has to be a close working relationship with the buying department. After all, it is the buying department that initiates the work for the VQA group by indicating that it would like to consider making use of a new vendor and would like to have the vendor assessed and graded. It is by no means inconvenient if the initial visit paid to prospective new vendors is a joint one by a VQA surveyor and a buyer. The surveyor will concentrate on the technical aspects, while the buyer is concerned with the commercial side. In this respect it needs to be remembered that the surveyor's eventual grading takes into account technical capability only, and that this is the one that matters most.

It is quite possible that following such a joint visit, the surveyor could give the highest grading to the vendor, while the buyer, from the commercial point of view, might give a "thumbs down." Decisions the other way around are equally possible, but it must be

VENDOR QUALITY ASSURANCE, LIAISON, AND BACKUP 163

accepted that if either the surveyor or the buyer says "no," that vendor will not be accepted no matter how high the grade from the partner.

It is also well to remember that some assessments about technical capability may well be on a restricted basis. Some companies manufacture a wide range of products, and it is not unreasonable for an assessment to be based on the vendor's capability to produce and control the quality of a single product line only. All others can be disregarded. If this is the outcome of an assessment, it will mean, among other things, that if the same vendor is to be considered later for another product line, he will have to be reassessed. This is, however, a contentious point with industry, and there are differing views about it.

In general, however, it is better to adopt the alternative scenario, that in which the vendor is assessed against his total quality control system for the complete product range. When this is done, it could be the case that the vendor's system for the control of quality will be considered generally good but found wanting in the one product line in which the surveyor is interested. In the long term this is certainly a much better way in which to carry out assessments, partly because it parallels the assessment methods used by defense establishments and many other large buying organizations.

Once the assessment and grading stages have been completed, it would be expected that a buyer would take the gradings into account when considering the final selection of a vendor on whom to place a particular order from among those who could supply the item to be ordered. Generally, when there is a choice, preference should be given to the vendor with the highest grading. This is because it means, or should mean, that the vendors with the higher gradings will be those whose products will cause fewer problems and fewer LOST COSTS to the customer after delivery from a nonconformance point of view. This is a factor that must receive most serious consideration and should be taken into account when price differentials exist. It must also be recognized that in any instance in which a buyer proposes to pass over a higher-graded vendor in favor of one with a lower grading, the buyer must expect to have to justify the decision, on sound commercial grounds, to the surveyor concerned.

However, when an order is to be placed with a new vendor, or for a new part from an existing vendor, or in any other case in which the surveyor considers it necessary, the buyer must inform the vendor of a number of things. In particular, these relate to the requirement for the submission of initial production samples from the tools. These samples will be closely examined by the various laboratories as may be necessary, and approved before production deliveries will be authorized. This is an important aspect of the buying role and is at the heart of the cooperation between the two departments. In this way the surveyor is made aware of the fact that a new vendor is being used or that something new is going to come from an existing vendor. When the surveyor knows this, he or she will be able to liaise with the vendor to the extent considered necessary. The surveyor will also be informed of repeat orders placed on existing vendors so that he or she knows the complete situation with respect to each vendor.

Of course, communication is, or should be, two-way. As the buyer keeps the surveyor informed, so the surveyor keeps the appropriate buyer informed of the current position with regard to individual vendors and their quality performance. Especially must the surveyor ensure that, when vendor ratings are worked out, the buyers are informed of the quality ratings which are determined for individual vendors on the basis of their actual quality performance. Once this information becomes available, which is not likely to be until a vendor has been supplying regularly for some months, it will be expected that the buyers will substitute the ratings for use as a target quality indicator in place of the original assessment grading. It must be remembered that the grading was a presupply subjective assessment of the vendor's expected quality capability while the quality rating is an objective determination based on the fact of actual performance.

Apart from quality ratings, the surveyor will advise buyers, routinely and in advance, of his program of regular surveillance visits to vendors and especially of specific problem visits to vendors. Apart from the general desirability of this cross-communication, individual buyers might occasionally like to accompany the surveyor, or in some cases, want the surveyor to take up a point with the vendor on the buyer's behalf.

When considering the general task of organizing a VQA group, there is one possibility that should be mentioned, the incorporation of the entire VQA group within the buying department. This may seem to be a somewhat revolutionary suggestion, although there were some rumblings in the buying profession in the late 1970s about the desirability of buyers having this responsibility under their direct control, that is, supplier assessment and control with full responsibility for the quality of all productive purchased material. There is much in this suggestion that is worthy of serious considerations, but there are a number of "if's." If such a proposition were to be seriously considered in a particular case, one would expect to see buying departments very different from most to be found today. For one thing, the department would have to be a very forward-looking, not at all wedded to the old ideas, as so many seem to be. There would have to be a very careful examination of a number of considerations. But there is one particularly important and major stumbling block to the actual implementation of such a proposal. It is most unlikely that it would be acceptable to the MOD and NATO under their Standards for Supplier Quality Assurance Systems. But the possibility of such control has at least been introduced and some of the problems that could arise have been, at least briefly, identified.

12
Selecting and Training Surveyors/Assessors

The nature of the task that vendor quality assurance surveyors are called upon to do is such that one must be sure of selecting good prospects. It is also a fairly new task in the field of quality and of such a degree of importance that it is essential that the right selection is made. It is worth going to a good deal of trouble to make sure that this is so.

One should first decide on the type of person required and prepare a job description. A possible list of the kind of qualities one might look for in a prospective surveyor is as follows: "presence," a good appearance, a confident approach, integrity, good "engineering" experience, broad quality experience, an understanding of factory costing procedure. VQA surveyors will be meeting, and perhaps discussing contracts with, managing directors at one end of the scale and details of individual specifications with shop operators at the other. They will be required to deal with a wide variety of products and processes, and although they cannot be expected to be a master of all, they should be able to pick out the main points and problems. A "plus" factor would be experience in incoming quality sections in a company that buys a wide range of parts for building into its own product. A good alternative to that would be experience in a company with many customers and having to deal with the customer quality problems that inevitably arise in such a case.

For English-speaking companies in a regional economic group such as the Common Market, there is one other important factor that is unfortunately considered too seldomly: the ability to speak fluently at least one other language. There is a commonly held belief for example, that most mainland Europeans speak English, but this is not true. In most countries, too few people take the trouble to learn a foreign language.

At this point a word of clarification is needed. This is with regard to the difference in meaning (if any) between the terms "surveyors" and "assessors" as used here. Whatever else they may also do, a prime task of assessors is in dealing with initial and recurring assessments of vendors to satisfy the assessor's principals that a vendor meets, and continues to meet, the basic quality control system requirements that are called for by whichever of the various standards discussed in Chapter 4 and 5 apply.

Although assessors may operate as a one-person team, the leader of a team of two or more people must be led by a registered lead assessor (see Appendix 1 of this chapter). It must also be explained that the term "principals" is used because a considerable number of assessments are undertaken by third-party organizations which will carry out the task on behalf of a purchasing organization; organizations which, for various reasons, either cannot or do not wish to carry out assessments for themselves. Surveyors, on the other hand, whether or not they carry out assessments tasks, will carry out for their own employing company the regular surveillance, audit, and other tasks described in Chapter 9.

It will be noted that there is no reference to experience in VQA. At the time of the creation of the group upon which these chapters about VQA are based, persons experienced in those techniques were very scarce indeed. It was much easier and better to look for good prospects and provide them with on-the-job training in the particular techniques that are needed. The order in which the desired qualities have been given is deliberate so as to emphasize those personal aspects that are likely to be of greatest importance in a job like this. However, at the time of the writing of this book, the shortage of experienced surveyors/assessors is being rapidly reduced as a result of the setting up and activities of the Lead Assessor Certification Board. This is discussed later in this chapter.

The next thing to consider, having selected surveyors, is training,

SELECTING AND TRAINING SURVEYORS/ASSESSORS

which will not be too easy. Many large buying organizations have introduced a variety of kinds of VQA, the British Ministry of Defence, (MOD) undoubtedly being the largest of these. However, with the exception of the MOD who might be regarded as a special case, there did not seem to have been any standardized, or readily available, method of training until now. Courses have been run for a number of years by the Portsmouth Management College, in the south of England, on behalf of the MOD. It is understood that about 10 of these courses take place each year and that they last for one week on a residential basis. Although they are primarily for MOD personnel, a few places on each have been offered to industrial companies on an invitation basis. In this way it is possible for both sides of the assessment problem to be displayed and discussed in a manner that would not be reasonably possible in any other way. This is because, in the main, it is probable that most of the industry representatives taking those courses will be "the assessed" rather than the assessors.

A considerable part of the course time is taken up with syndicate working on case studies. During those syndicate sessions, judgements are made on the basis of the descriptions given of situations in a hypothetical but otherwise realistic industrial company. In passing it should be mentioned that training using the participative method with syndicates working on a case study is by far the most successful and efficient method of training. The quality of these courses is such that they have also been run by Portsmouth College in Holland, for the Dutch MOD.

Clearly, there will not be many purchasing organizations that can undertake the task of running their own comprehensive training courses as is done by the Portsmouth College for the MOD. It must also be remembered that if a purchasing organization of any size wishes to carry out all the assessments that it wants to have undertaken and to use its own personnel exclusively for the purpose, it may feel that the need for training is not so important. However, training will be necessary if for no other reason than to ensure that the organization's assessors/surveyors work to an acceptable and consistent standard, with the emphasis on "consistent." This is most important, particularly as it is probable that many of the vendors being assessed will already have been assessed by other organizations. If a vendor feels that there has been a significant difference in standards of assessments, and unfortunately there are at times, there will be problems for all parties.

However, should a company wish to introduce its own training program for surveyors or assessors, the basic requirements should be the following as a minimum:

1. There should be a formal and documented training program based strictly on the requirements of BS 5750 (see Appendix A) or other appropriate and equivalent standard (see Chapter 5).
2. At least 50% of the course content should consist of syndicate work on case studies of real or realistic situations.
3. A high and uniform program standard is set.
4. Some form of certification of capability is set and it is recommended that a pass level of at least 60% be set.

Until quite recently there were no formal training courses for surveyors/assessors other than the MOD course. But now the situation has changed. For some time there had been a realization of the existence of this problem and of the urgent need for a solution to be found. In the United Kingdom the initiative in this respect was taken by the Institute of Quality Assurance (IQA). After much preliminary negotiation, and with the blessing of the Department of Trade and Industry, with secretarial services being provided by the IQA, a Lead Assessor Certification Board has been set up. It is now a requirement that assessments be carried out by a team working under a lead assessor (upon occasion the "team" will consist of a lead assessor working alone).

Lead assessors must be approved by this board. Specific experience and training requirements are stipulated. The problem that arises when one attempts to set a standard for surveyors, especially obtaining a form of certification as an indication that the standard has been achieved, has been difficult. However, progress has been made to such an extent that a set of requirements has been formulated and agreed to for assessment team leaders. For most industrial companies this could be taken to mean senior assessors or surveyors.

As a result of the creation of the Lead Assessor Certification Board, which has been operating since about mid-1985, a number of residential courses have been approved by the board as being of the required standard. Since that time the board has been receiv-

SELECTING AND TRAINING SURVEYORS/ASSESSORS

ing and approving application from many people for registration as lead assessors.

Persons training as VQA surveyors are presenting an image of their companies, so it is most desirable, for many and obvious reasons, that they present the best possible image. This applies whether they are working for a purchasing organization or company or for a third party assessing organization, in which case it is even more important to present a good image.

In Appendix 1 the "Qualification and Experience Requirements" and the "Code of Conduct" for lead assessors are presented by permission of the Institute of Quality Assurance.

APPENDIX 1: EXTRACTS FROM THE "RULES AND PROCEDURES OF THE LEAD ASSESSOR CERTIFICATION SCHEME"

Qualification and Experience Requirements for Lead Assessors

1. Organizations or individuals applying for registration of Lead Assessors, shall be able to demonstrate that nominees meet the following requirements:
   (a) They should be academically qualified in both a recognised scientific/technological discipline (eg. Engineering, Pharmacy, Chemistry, etc.) and in quality assurance. For guidance the following is a non-exclusive list of acceptable qualifications:
   Scientific/technological: a suitable degree, C. Eng. or HNC/HND
   Quality assurance: the academic level and practical experience required for Corporate Membership of the Institute of Quality Assurance
   (b) Have satisfactorily completed a training course in quality assurance management system assessment recognized by the Management Board and have satisfied any examination or test of competence embodied in the course.
   (c) Have had a minimum of five years recent and relevant practical experience, of which at least three years have

been devoted to the application of quality assurance principles.
(d) Have taken part within three years immediately prior to application, as a member of a quality assurance management system assessment team, in assessments of at least five companies, to the satisfaction of the assessment organization. Such assessments must have been to BS 5750 Parts 1 or 2, or the equivalent standards of a major purchasing organization.
(e) Have competent working knowledge of nationally recognized quality assurance management system assessment standards.
(f) Have the necessary personal qualities of integrity, tact and character to perform the duties of a Lead Assessor.
(g) Have confirmed willingness to accept and abide by the Code of Conduct laid down by the Institute of Quality Assurance and set out below.

Code of Conduct for Quality Assurance Management Systems Lead Assessors

Before registration as a Lead Assessor each candidate is required to confirm willingness to observe and be bound by the following code:

(a) Shall act in a strictly trustworthy and unbiased manner in relation to both his/her employer (the assessment organization) and any companies involved in an assessment by him/her or personnel for whom he/she is responsible:
(b) Shall not accept payment, gift, commission, discount or any other profit from the companies assessed, from their representatives, or from any other interested person nor knowingly allow personnel for whom he/she is responsible to do so:
(c) Shall disclose to his/her employer any relationships he/she may have with the company to be assessed before undertaking any assessment function of that company:
(d) Excluding discussions essential to the efficient conduct of the assessment, shall not disclose the findings, or any part of them, of the assessment team for which he/she is responsible, to the company being assessed, or to any third

SELECTING AND TRAINING SURVEYORS/ASSESSORS 173

      party, unless authorized in writing by the assessment organization to do so:
      N.B. It is the responsibility of the assessment organization to obtain the permission of the company before communicating any part of the findings to a third party.

(e) Shall not act in any way prejudicial to the reputation or interest of the assessment organization, or to the assessed company:

(f) Shall, in the event of any alleged breach of this code, be subject to a formal enquiry procedure operated by the assessment organization.

13
Record Systems

It may seem to be the case, when one first considers the amount and variety of the various types of records and documents that have to be developed and maintained for the effective functioning of a VQA/BOQA operation, that it is all too much! But a more mature reflection will bring the following realizations: first, that a VQA operation can produce results that are economically rewarding to the company; second, that the records are not as voluminous as was at first thought; and third, that they are helpful in a number of ways. I am, at this point, reminded of the comment of the executive vice-president of a large multinational American company when he was showing an important customer around the company's main factory in Dallas when the visitor remarked upon the large number of bulky computer printouts which he saw throughout the factory. "Yes", said the vice-president, "there are a lot of computer printouts/paperwork all over the plant, but let there be no mistake about it, this company has got where it is *because of*, not in spite of all that paper."

For a long time it has been common practice for record systems in quality departments to be developed in written form in a variety of ways. Increasingly today, however, it is the case that more use is being made of computers for this purpose. Generally, because of the cost of computers and the complexity of the programming required, it is the larger companies that have been able to turn over their systems of records to computers. However, with the constantly decreasing cost of computer systems and the simplification of methods of programming, it is now possible for small companies to consider their use, especially when small systems suitable for small companies can now be bought for only a few hundreds of dollars. But a word of warning is necessary.

One reads many glowing reports these days about the performance and capabilities of small computer systems, especially the word processor and mini-or microcomputer. But there are also many cases in which the expectations of owners of small (and not so small) businesses as to the benefits of systems they have bought have been dashed, either because the equipment would not do the things that were claimed for it, or because the instructions for using the software were not readily understandable, which produces the same lack of results. There are many reasons for this; let's look at one or two of them, at least.

The first is one that can be of particular importance to those who many wish to buy a small computer system. That is the "enthusiasm" of sales and/or marketing personnel, who will make extravagant claims for the capabilities and versatility of the equipment they are trying to sell which are not met in practice. A second reason, also of importance, is the poor quality of the instructions contained in the manuals supplied with the systems.

Small companies should make themselves very clear about the tasks they wish a computer system to do for them, then get the salesperson to demonstrate that the system he or she is trying to sell them will do what the customers want it to do.

Let's get back to record systems in general. In discussing systems for the records of a BOQA and a VQA group, one must realize that there is an option of three approaches:

1. Use the computer all the way.
2. Start with written records, with the thought of eventual switchover to computers.
3. Start with written records with no intention of changing over later to computerized records.

I would recommend, particularly for a smaller company, that the second option be chosen with the third as second choice. This is because when new systems of this nature are being created there will always be problems. These problems almost certainly will

RECORD SYSTEMS

necessitate changes in the system, and it is easier to change a system of initial written records than it is to change a computer program. There is also a smaller staff training problem with written records.

No attempt will be made here to discuss in further detail the use of computerized records, or changing over to them from written records, because of the specialized systems analysis and programming skills that are required. Discussion will be confined to the second and third possibilities, which will cover most cases. Should a company wish to proceed in the direction of computerized systems, it would be well advised to seek those special skills from the beginning of planning. The system of records discussed next is basically that used by Company D.

There are two main groupings of records that need to be considered, although there may be others that could also be considered. These two groups are the ones directly associated with the work of VQA surveyors: probably by product type grouping or by geographical location of the vendors. In either case the main identification will be the names of the individual parts and their vendors. Each surveyor maintains his or her own files. In the file should be the complete history of each vendor's association with the company, beginning with a copy of the original letter requesting the initial appointment for the VQA surveyor to visit that vendor. In chronological order this should be followed by the assessment records and grading decisions and the report and results of routine surveillance visits. In cases in which there had been problems, one would expect to find all the correspondence relating to problems. It is also suggested that careful consideration be given to the following point about vendor "name" files.

From time to time a particular problem that arises with a vendor will generate a considerable amount of correspondence and reports. When this happens it is an excellent idea to set up a subsidiary file for the vendor, with the problem title as its subheading. This file will contain all the data for the particular problem. The main vendor file should contain a short "aide-memoire" note, located chronologically, with a reference to the subsidiary problem file, giving its title.

The bought-out quality audit (BOQA) records will be much more extensive than those of the surveyors and it is primarily here that any consideration about eventual computerization should be given. Unlike the surveyor records, which are largely correspondence and much less amenable to computerization, the BOQA records are more suitable for eventual computerization. Just about all the data which they contain can be expressed and recorded in a manner that is acceptable for use by computer, that is, by groupings of letters or numbers or a mixture of both. The main classification will probably be by part number and then by part vendor. A subsidiary and cross-referenced classification will be by vendor name and then by all the part identifications supplied by that vendor. This secondary classification is primarily for reference purposes but is also of use for vendor rating purposes. It will not normally contain the detailed information about inspection results, concessions, or material disposition that will be contained in the main records.

Basically, there must be for each part, and for each vendor of that part, a record card. On that card will be entered the "shorthand" results of the inspections that are carried out on individual shipments of that item from that vendor. As the card is the basic information source about the item, it must also have provision for adding any necessary information about concessions. It is also required for the guidance of the audit inspector in determining the appropriate sampling plan and level of inspection to be used in each case. During their routine scrutiny of the record cards relating to "their" vendors, there must also be a suitable "signalling" arrangement for informing the surveyors when a problem situation has arisen or is considered to be likely.

With regard to concessions, the record must be capable of distinguishing between those concessions that are applied for internally, because there is a need to use material that would otherwise be rejected, and cases in which the concessions are requested by the vendors. This separation is essential because it is important that no penalty accrue to a shipment that is submitted under cover of a vendor-requested concession. This point is discussed more fully later, always provided of course, that the shipment was not rejected for some other fault.

A further important factor, which is based on information taken from the basic record card, is that of vendor rating. This subject was mentioned in Chapter 5 and 6 and discussed fully in Chapter 7. Except that the vendor quality ratings are determined from the

RECORD SYSTEMS 179

basic record cards, and on a regular basis, no more needs to be said about it here.

In the main, discussion has proceeded on the basis that written records are in use with the data entries made by the audit inspectors directly onto the cards. Similarly, the record examinations made by the surveyors are conducted entirely on the basis of a visual examination of the cards, during which they are looking for problem signals. If, however, a computer system is in use or envisaged, the techniques used will be rather different. Data recording, recall, and examination will be via the medium of video terminals.

When a computer system is used and an audit inspector wishes to begin inspection of a received shipment, he or she will "key in" the appropriate code for the part and the vendor. The inspector will then see, on the video screen all the information that is needed to carry out the inspection. This will include details about the sampling plan, inspection levels, and so on. (The inspector will still, of course, require the audit specification for the part.) Upon completion of the inspection, the inspector will "key in" the results, while watching the simultaneous screen display for "keying in" errors, and then confirm it, when it will be added to computer memory.

Surveyor scrutinies will clearly have to be conducted on a different basis: probably by having the surveyors key in a prearranged code whereupon the screen will display in sequence any possible problem situations. These can be retained on the screen for examination or cancelled for a following display at a signal from the surveyor. Suitable arrangements should be made so that printouts can be supplied if it is desired to reproduce any display from the screen in a more permanent form for later, more leisurely examination. A degree of control will be needed for such a facility, as it is a relatively expensive addition to a computer system.

However, it must be appreciated that with a normal card system all calculations, including those for a vendor rating system, will have to be hand done as required, although one may be sure that hand calculators will be used for all the arithmetical calculations. In this way a "running" vendor quality rating is possible. The reverse is the case with a computer as all calculations, for whatever purpose, are done automatically. This does mean that for any part and vendor a current vendor rating figure can be obtained at any time by asking the right question of the computer. Of course, from the card system the same facility is available. It may just take a little longer to work out each time.

When setting up a VQA group it is possible to get it going without doing anything beforehand about the preparation of the various standard forms that will eventually be essential for the efficient functioning of the group. However, this policy is not recommended. It is well worthwhile spending time at an early stage, perhaps during the initial personnel recruitment stage, on the development of the documentation system that will be required. Even internal recruitment of personnel will take time—enough for this task.

Quite a number and variety of "standard" forms will be required for efficient operation. A typical list follows and each one will be discussed individually. It must be stressed that this list is not necessarily the ideal one, but simply as a basis for consideration and development in other circumstances. Circumstances are bound to vary from company to company and requirements will also vary accordingly.

1. Standard letter to vendor managing director requesting an appointment for the VQA surveyor's initial visit
2. Preliminary assessment visit report and grading record
3. Vendor notification of assessment and grading
4. Initial production sample approval form
5. Standard letter accepting initial production samples
6. Standard letter rejecting initial production samples
7. Vendor concession application form
8. Vendor request for design change form

1. Standard letter. Ordinarily, this letter will be sent by the VQA manager to explain briefly the purpose of the visit and the policy of the company. Accompanying it should be a copy of the vendor quality assurance manual, if there is one, so that the vendor has an opportunity to study the requirements in advance. The vendor will then, it is hoped, be fully prepared for the surveyor's visit, with a list of questions about the vendor's place in the scheme of things and how the company will be affected.

2. Preliminary assessment visit report and grading record. This will usually be in the form of a standard questionnaire that will be completed by the surveyor after he or she has made the initial visit. It will be used as a basis for planning the formal assessment that will later be made to determine the degree of comformance, or otherwise, to the full requirements of the quality management systems standard used, BS 5750 or equivalent, and to determine the appropriate level of the standard that should be applied. This point, of course, would have been discussed by the surveyor with the vendor's staff. If the vendor has a quality manual, a copy of it will also be requested and taken back by the surveyor.

3. Vendor notification. Although the vendor should have been informed of the outcome of the assessment visit at the time on an informal basis, this is a standard letter that formally advises the vendor of the outcome. It is also required for record purposes, for both the vendor and for the surveyor's own files and those of the buying department. For the buying department, it is an important piece of information.

4. Initial production sample approval form. This is a four-part form that has to be completed by the vendor when he or she supplies the initial production samples for approval. The various copies of the form are used to initiate the various examinations that will be made of the samples. These examinations will include dimensional measurements and/or laboratory examinations, as may be appropriate. They are intended to assure the company that the vendor's tooling is capable of volume production to requirements before production deliveries take place, to confirm that his or her measurement capability is satisfactory as well as any particular treatments that may be required. At this stage it is relatively easy to correct any deviations that may have occurred before bulk shipments have been produced and sent. These samples will normally be required for all orders for parts not ordered previously from the vendor; whenever additional sets of tooling are brought into use, as in the case of multiple-impression tools; when manufacturing methods, or materials, change; or whenever the customer feels that further samples are necessary.

5. Standard letter accepting samples. This letter formally advises the vendor that his or her samples have been found to be in conformance with requirements and that production deliveries may now begin in accordance with the schedules given to him or her by the buying department.

6. Standard letter rejecting samples. This letter formally advises the vendor that his or her samples have been found not to be in conformance with requirements for the reasons stated and are therefore rejected. The letter also requests that the vendor submit a further set of samples for approval, and a new set of forms for this purpose will be enclosed. The vendor is reminded that he or she must not begin production deliveries until he or she has received formal advice that the samples have been accepted.

7. Vendor concession application form. There are two particular points that should be borne in mind when considering this form. The first is that the customer does not want to receive unexpected and unsuspected nonconforming material; It will cause a variety of problems. The second is that the vendor must be encouraged to follow this line so that he or she will not try to "slip in" nonconforming material in the hope that it will not be discovered. It has to be recognized that there will be occasions when defective material will be produced. When that does occur, it is a positive sign if the vendor considers the parts to be usable and submits a request for a concession. In such an event, if the concession is granted, the shipment should be regarded as an accepted shipment—provided, of course, that there is nothing else wrong with it. On the other hand, if it is found that it is not possible to accept the shipment, it should not be recorded and certainly not regarded as a rejected shipment.

8. Vendor request for design change. It is a fact of life that no person knows everything. Therefore, it may be that occasions will arise when it will be found that a vendor knows more about an item ordered than the customer does. It may be a simpler or cheaper part for a particular function. It may be an alternative method of surface treatment or different materials or, even, a different part altogether. It may also be just because the vendor is not able to meet the specified requirements with the available equipment and/or methods. Or it may even be that the stated requirements are at the boundaries of knowledge of manufacturing practicality.

Whatever the reason, there will be occasions when the vendor

RECORD SYSTEMS

will want to be able to do things differently and not according to the requirements. In those circumstances the vendor will ask for a change to be made to the design to suit their proposals, and this is the form the vendor will be required to use for the purpose. Apart from the usual standard information for record purposes, the vendor will have to explain fully the change he is requesting and, most important, the reason for requesting the change. When received, the completed form will ordinarily be dealt with by the design department. It is not a form that will automatically be supplied to the vendor but one that the vendor has been told about and which is available on request.

Summary of the Detail About the Forms

The standard letters and forms that have been described make up a basis for the documentation for a VQA group operation. Of course, in different circumstances there may well be a need for alternative documents, but it should not prove difficult to prepare an alternative set of documents for those circumstances.

As in the case of some earlier chapters, copies of five real standard procedures have been included in appendixes of this chapter. They have been included, as before, for the benefit of readers who wish to delve a little more deeply into some of the practicalities of documentation needs when dealing with vendors. These are operational procedures whose purpose is different from that of the set of standard forms described earlier in this chapter. They were used by Company A and their titles are as follows:

1. Concessions and Production Permits
2. Requirements for Subcontractors Records
3. Inspection Documentation and Records for Goods Inwards Inspection
4. Purchasing Documentation
5. Retention and Disposal of Records

As before, a few deletions and slight changes have been made to avoid identifying the company.

An explanation may be useful of the difference in meaning of the terms "concession" and "production permit."

> *Concession.* In general, a concession would be requested when something has been produced to requirements incorrectly but

it is thought that it could be used if consideration is given to the nature of the nonconformance.

Production permit. If, for some reason, it is considered to be either desirable or necessary to produce something not in accordance with the requirements, a production permit would be requested before manufacture begins. Ordinarily, a production permit would be granted for a specific quantity or period of time.

APPENDIX 1: PROCEDURE FOR CONCESSIONS AND PRODUCTION PERMITS

1. Purpose of the Procedure. To set out the situations in which applications for concessions and production permits can be applied for and the method of making such applications and distinguishing between the three types of application forms which can be used according to the circumstances.

2. Scope. The two manufacturing divisions of the Company and the quality assurance department are involved. The first, generally, as originators of applications and the second as the processors, and decision takers where indicated, of the applications within the Company.

3. Related and Relevant Procedures and Other Documents

Def. Stan. 05-21, paragraph 215

4. Responsibilities

4.1. The prime responsibility rests with the manufacturing divisions for raising applications for concessions or production permits when the appropriate circumstances arise.

4.2. The responsibility of the QA department begins after an application has been made, to ensure that it is properly routed to the appropriate authorities and that a follow-up is maintained to ensure completion within a reasonable time period.

5. Requirements of the Procedure

5.1. *Definitions*

5.1.1. *Concession*

5.1.1.1. A concession is required when a product has been made incorrectly to specification and it is considered that perhaps it could be used, or must be used. It may also be the case that before it could be used, certain additional steps, or precautions, might have to be incorporated.

5.1.2. *Production Permit*

5.1.2.1. A requirement for a production permit can only arise *before* the product has been made (e.g., it may be specified that a product has to be made from a particular material). The material is not available and it is proposed to make the product in another material, which, it is considered, will equally, or sufficiently nearly so, meet the requirements of the original specification.

5.2. *Concession Classes*

5.2.1. There are, for Company purposes, three classes of concessions, which are categorized in the following paragraphs (the same applies to production permits).

5.2.1.1. *Class A*

5.2.1.1.1. This class is the highest and of the most importance.

5.2.1.1.2. This class includes all those cases in which the deviation from the specification in the product is such that it will, or may be considered likely to, affect one or more of the following factors:

Safety
Functioning
Strength
Interchangeability
Product life
Maintainability

5.2.1.2. *Class B*

5.2.1.2.1. This class includes those cases which, while not falling into class A, are considered to be of sufficient importance to ensure that a formal permanent record is kept of them and, also, when it is considered advisable for the advice or guidance of the design authority to be sought.

5.2.1.3. *Class C*

5.2.1.3.1. This class is the lowest level and will include all those cases that do not fall into either of the two higher classes. The deviations in these cases are considered to be minor and may be dealt with by the appropriate QA engineer without any need for more formal concession application action. All these class C concessions will be recorded in a record book maintained by the QA department.

5.2.1.3.2. From time to time an examination of this record book may be requested by a representative of the design authority for the product or products concerned in any class C concession. Following upon such examination the representative will "sign off" the book immediately after the most recent class C concession entry as an indication of general acceptance of the classification and the actions that have been taken. If the representative questions any entry or action, he or she must take the matter up with the quality manager and any following actions will be dependent on those discussions.

5.3. *Concession Applications*

  5.3.1. *The Forms*

    5.3.1.1. There are three forms, depending on particular circumstances, which may be used. These forms and the circumstances in which they are used are described in the following paragraphs.

  5.3.2. *Ministry of Defence Production Contracts*

    5.3.2.1. The form is MOD Form 77.

    5.3.2.2. The requestor of the concession will complete seven copies of the form [copies can be obtained from the inspection record office (IRO)] and fully answer questions 1 to 13 in part 1 of the form and also "sign off" and date, etc., the small section at the foot of the front of the form. The requestor will then pass the seven copies of the form to the relevant inspection supervisor.

    5.3.2.3. The relevant inspection supervisor will check and agree to the details and initial the form beside the requestor's signature at the bottom left hand corner of the form and then pass all seven copies to the IRO.

    5.3.2.4. The IRO will register the application in their concession application register and allocate it the next consecutive serial number.

      5.3.2.4.1. The IRO will file one copy of the application and pass the remaining six copies to the MOD quality assurance representative for transmission to the design authority for a decision.

RECORD SYSTEMS

5.3.2.4.2. While the application is with the MOD the IRO will monitor the situation and will follow up as necessary.

5.3.2.5. Upon eventual receipt of the decision of the design authority, which will be detailed on one copy of the original set of application forms, the IRO will pass it, via the chief inspector for his or her information and possible additional action, to the inspection supervisor who checked and agreed the original application.

5.3.2.5.1. The inspection supervisor will take such actions as are required following from the design authority decision, in conjunction with the relevant production supervisor and the QA engineer, and will suitably record those actions. Upon completion of those actions the inspection supervisor will return the copy of the application form to the IRO for final filing in their "completed concession" file.

5.3.2.6. The IRO will file the completed form.

5.3.3. *MOD Experimental and/or Design Contracts*

5.3.3.1. The form is I.Arm. form 629.

5.3.3.2. The occasions on which this form is likely to be used will be few and far between. It is not anticipated that an occasion will arise in the foreseeable future. However, should such an occasion arise, appropriate steps, in cooperation with the MOD quality assurance representative, should be taken to process the form.

5.3.4. *Non-MOD Contracts*

5.3.4.1. The form is Company form ADMEL 0999/1.

5.3.4.2. Except that this form is a three part NCR (no carbon required) form, the procedure to be used in preparing an application is as in paragraphs 5.3.2.2. to 5.3.2.4. inclusive. The forms are serially numbered so that there is no need for the IRO to allocate a serial number to them.

5.3.4.2.1. The IRO will pass the top copy to the Company engineering department (ED) in all cases.

5.3.4.3. The ED will classify them as Company design responsibility or non-Company responsibility and deal with them as follows:

5.3.4.3.1. Those that are company design responsibility will be dealt with by the ED itself and returned to the IRO with the decision as in paragraphs 5.3.2.5.

5.3.4.3.2. Those that are "special" design responsibility will be passed over to the "special" division for action. In due course, as the "special" division decision is received, the ED will take note of the decision and again pass the copy of the form to the IRO as in paragraphs 5.3.2.5.

5.3.4.4. The IRO, on receipt of copy forms with recorded decision, will deal with the forms as in paragraphs 5.3.2.5, passing it to the inspection supervisor concerned.

5.3.4.4.1. The inspection supervisor will then act as in paragraph 5.3.2.5.1. and return the copy form, eventually, to the IRO.

5.3.4.5. The IRO will file the completed form.

5.4. *Class "C" or Minor Concessions*

5.4.1. Records of these minor concessions will be kept in a "minor concession" book. There may be one of these books in each inspection area of the QA department. Their number and location will be decided by the quality manager.

5.4.2. Responsibility for the holding and safe custody of each book will rest with the supervisor of the designated inspection area. In the first instance it will be his or her responsibility to ensure that the full formal details of the deviation and concession are entered into the book. Of course, other normal identifying data, such as the date, will also be entered into the record.

5.4.3. The inspection supervisor will ensure that the request for a minor concession is brought to the attention of the relevant QAE as quickly as possible so that he or she may make a suitable decision.

5.4.4. The actions that will subsequently be taken will depend on, and be in accordance with, the decision of the QAE and will also be recorded in the minor concession book against the request and its details.

5.5. *Production Permit Applications*

5.5.1. *The Forms*

5.5.1.1. The forms are the same as are used for concessions, but the word "concession" is deleted and the words "production permit" are added.

5.5.2. The completion of applications and the general procedure for handling the forms, and decision taking, is exactly the same as has been described for the various classes of concessions.

APPENDIX 2: PROCEDURE FOR REQUIREMENTS FOR SUBCONTRACTOR RECORDS

1. Purpose of the Procedure. To set out the requirements expected of suppliers to the Company about the records they must keep.

2. Scope. The supplier control engineer (SCE) of the quality assurance department and the purchasing department are equally concerned and there may also be an involvement on the part of the subcontract controller.

3. Related and Relevant Procedures and Other Documents

Def, Stan. 05-21, paragraph 205
03-702-001   Assessment, Selection, and Control of Subcontractors
03-702-002   Evaluating a Subcontractor
03-713-001   Release of Company "Free Issue" Material to Suppliers
06-101-002   Retention and Control of Records

4. Responsibilities. Although the primary responsibility is that of the purchasing department, the method of working is such that the actual exercise of responsibility must rest with the SCE who actually approves the suppliers and maintains surveillance over them.

5. Requirements of the Procedure

5.1. The records that have to be prepared by the supplier do not have to be held by the Company, even although they are regarded as an extension of the Company records. They should normally be kept by the supplier but be made available as necessary for examination by Company QA personnel and the MOD quality assurance representative (QAR) as they may think fit.

5.2. The same conditions apply to the retention and destruction of these records as apply to Company records. Thus they may not be destroyed or disposed of in any way without first seeking the approval of the Company in writing. Such approval will be given only if the Company is satisfied that such disposal will not nullify the Company's own records of which they are an extension, and if the QAR will also agree.

5.3. The types of records that the supplier is required to prepare and maintain are described in the following paragraphs:

5.3.1. The records fall into two broad categories, of which the first may be said to be directly operational records. (The second category is described in paragraph 5.3.2.). These are the records that are made and kept during actual production and inspections carried out on the material being worked on by the vendor.

5.3.1.1. Those records include batch or route cards or other documents that identify batches of materiel and identify the operations that are to be, or have been, carried out on them, together with the results.

5.3.1.2. Those records also include the records made of the various inspections that are carried out during the course of the production cycle. They may include the results of first-off and patrol inspections, as well as of final inspections, and will include details of the dispositions of the rejected parts, including scrapping if that was the outcome.

5.3.2. The second category of records might be described as indirect. These are the records of functions that support the actual manufacturing operations.

5.3.2.1. Included in these are the records the supplier maintains of the gauges and measuring equipment used on Company work. In the event that all tools and measuring equipment have been supplied on loan by the Company, it is still a responsibility of the supplier to assure him or herself that these items are being cared for and that they take suitable steps to ensure that the equipment is not allowed to "drift" outside specified tolerances for use by means of suitable calibration procedures.

5.3.2.2. In those instances in which the supplier is carrying out processing operations for the Company (any such processing operations will be permitted *only* by the first-obtained written joint authority of the Company's quality manager and chief metallurgist), they must ensure that the records they maintain of the operation and control of these processes are to the requirements of, and approved by, the Company chief metallurgist.

5.3.2.3. As in the case of "direct" records, these records shall provide for clear identification of any instance in which a process goes out of control and also of the resultant action on any material that might be undergoing the process at the time. The records will also ensure that details of any corrective action taken to bring the process back into control are recorded with identification of the persons directly concerned.

APPENDIX 3: PROCEDURE FOR INSPECTION DOCUMENTATION AND RECORDS FOR GOODS INWARDS INSPECTION

1. Purpose of the Procedure. This procedure is intended to identify the various forms and documents that are used and referred to during the course of the inspection of incoming materiel within the goods inwards inspection (GII) area of the quality assurance department.

2. Scope. This procedure is applicable, directly, to the GII area, but there is a relevance to other areas because of the transit of paperwork between different areas.

3. Related and Relevant Procedures and Other Documents

Def. Stan. 05-21, paragraph 208
05-202-001   General Procedure Governing Incoming Inspection

4. Responsibilities. Direct responsibility for the correct functioning of this procedure rests with the supervisor and inspectors in the GII area. It is their task to ensure that the various forms are correctly completed, distributed, and filed in accordance with the stated requirements.

5. Description and Routing of Forms

5.1. The various forms are described in the approximate order in which they have to be dealt with in following procedure 05-202-001 in the GII area.

5.2. *The Forms*

5.2.1. *Goods Receiving Note*

5.2.1.1. A set of these forms is raised by the goods inwards department for every shipment of material that is received. Four copies (blue, yellow, white, and pink) are sent with the shipment to the GII area. Upon completion of the inspection, the yellow copy, suitably endorsed with the inspection results (as are all the others), stays with the material and the remaining three are distributed as follows:

> The white copy goes to the original requisitioner.
> The pink copy goes, via the inspection records office, to the accounts department.
> The blue copy is filed in the GII area files.

5.2.2. *Incoming Inspection Instruction (P1053)*

5.2.2.1. This instruction is prepared by the supplier quality assurance engineer and is a detailed inspection instruction governing the inspection of a single item from a single supplier. These instructions are kept in the GII files.

5.2.3. *Release Notes or Test Certificates*

5.2.3.1. These are documents received from a supplier which certify that inspection and/or testing has been carried out in accordance with the Company's order instructions, or that materiel specifications are as detailed or as called for in Company orders. They are first received by the chief inspector and passed by him, in the first instance, to the GII area to assist in the inspection of incoming shipments. They are eventually returned by the GII area, upon completion of inspection of the relevant shipments, to the IRO endorsed with the GRN serial number and the laboratory MSD form serial number that relates to the particular shipment.

5.2.4. *Form R494*

5.2.4.1. When the results of a GII inspection are unsatisfactory on a particular shipment, the inspector will record the results on this one-page form.

5.2.5. *Bought-Out Inspection Report*

5.2.5.1. This is a four-part multicolored set of forms prepared by the IRO from form R494 made out in GII. It is distributed by the IRO to the supplier, the relevant purchasing section, and to shipping with the fourth, file, copy for themselves.

RECORD SYSTEMS 193

5.2.6. *MSD Form*

5.2.6.1. This form is prepared by the metallurgical laboratory to indicate to GII their requirements for test samples from particular shipments. It is sent to GII in response to requests from that area.

5.2.7. *Quality Report Diagnostic Chart (XR Chart P2083)*

5.2.7.1. This chart is prepared by GII in the period during which they are awaiting instructions from the metallurgical laboratory. It is used to record, in mean and range formats for subsequent analysis, specified important dimensions of individual items of high-volume parts.

5.2.8. *Inspection Control Plan*

5.2.8.1. This is a form basically similar to form P2083 insofar as its basic purpose is concerned. It is used to record full details of the inspection carried out on all-shipments other than bar, sheet, and high-volume parts.

5.2.9. *MSM Form*

5.2.9.1. This form is used by the metallurgical laboratory to record the results of satisfactory examinations and also to advise GII of the results.

5.2.10. *MSM Rejection Form*

5.2.10.1. This is similar to the MSM form except that the metallurgical laboratory uses this one to record the results of unsatisfactory examinations. Again it is used to inform GII of the results and also triggers further action by them to dispose of the shipment.

5.2.11. *Quality Action Request   Incoming Material (P1079)*

5.2.11.1. A single-part form prepared by GII to seek disposal instructions for shipments found to be dimensionally nonconforming at the GII inspection. It is routed via production control, engineering, planning, and the supplier control engineer back to the GII area with instructions regarding disposition and disposal of the shipment.

5.2.12. *Weekly Summary Sheet (P1077)*

5.2.12.1. Prepared on a weekly basis, this form is used to record in summary form the results on all shipments cleared during the week. It is distributed to the purchasing and production departments so that they may be informed of the current situation regarding incoming material.

APPENDIX 4: PROCEDURE FOR PURCHASING DOCUMENTATION

1. Purpose of the Procedure. To indicate the basis for and type of documentation that is to be kept by the purchasing department.

2. Scope. This procedure concerns the purchasing department alone.

3. Related and Relevant Procedures and Other Documents

Def. Stan. 05-21, paragraphs 208 and 210
03-701-001   Requirements for Subcontractor Records
03-711-001   Requirements for Subcontractor Gauge Control
03-718-001   Purchasing Responsibilities in the Quality System
03-718-003   Purchase Order Change Control

4. Responsibilities. The responsibilities alloted under this procedure are those of the purchasing department alone.

5. Requirements of the Procedure

5.1. The records that the purchasing department must keep cover a variety of requirements and formats. They are dealt with in the following paragraphs, but there is no significance about the order in which they are described. The list should not be taken as complete but is intended primarily to indicate the extent and scope of what is required.

5.2. As early as possible in the life of a contract the purchasing department (PD) should prepare a listing of the items and/or services that are to be purchased, and this list should become the basis of the complete purchasing programme for the contract.

5.2.1. The list should specify the names, drawing, and/or specification numbers for the items, with issue references for those drawings and/or specifications, and where and how objective evidence of suitability will be demonstrated.

5.2.2. The list should also provide the necessary recording facility for entering the relevant purchasing requisition numbers, order numbers, and details of documentation sent with orders, including drawings and/or specifications, etc., and issue references of them.

5.2.3. Requisitions and orders will be suitably filed and cross-referenced as may be necessary to relate to each other and/or the above-mentioned list.

RECORD SYSTEMS

5.2.4. In the case of proprietary items or equipment that have been specified by the design department, the PD will also record in a suitable place in the records the authority from the design department for purchasing the specific items that have been approved by them. (Care must be taken to ensure that when such items of proprietary equipment are being purchased, only approved items are purchased.)

5.3.5. The identity of suppliers from whom materiel or services have been ordered will be indicated on the list, and this indication may be by means of the Company approval number allocated to a supplier after the supplier has been approved by the Company.

APPENDIX 5: PROCEDURE FOR RETENTION AND DISPOSAL OF RECORDS

1. Purpose of Procedure. To set out the policy for the general retention and disposal of the essential records that are required to be kept in conformance to the requirements of Def. Stan. 05-21 [see Appendix A].

2. Scope. The procedure is applicable to all areas of the Company that are required to keep records by the requirements of Def. Stan. 05-21. These include, principally, the quality assurance, engineering and design, purchasing, and manufacturing departments.

3. Related and Relevant Procedures and Other Documents

Def. Stan. 05-21, paragraph 205

4. Responsibilities. Prime responsibility rests with the quality assurance manager and his or her department, but the managers of all the other departments mentioned also have responsibilities in ensuring that their departments comply fully with the requirements.

5. Requirements of the Procedure

5.1. The records that are the subject of this procedure are those "records that demonstrate the effective operation of his [the contractor's] quality control system; ... pertinent sub-contractor records are an element of this data."

5.1.1. This procedure specifically refers to the records pertaining to MOD contracts. However, there may be a similar restriction on records for non-MOD contracts. In the latter event, reference should be made to the quality assurance manager before disposing of any records produced by any of the departments, and sections of departments, listed in paragraph 5.2.

5.2. This means that the records maintained by the following departments and sections of departments are concerned.

> Engineering and design department
> Planning and production control departments
> Purchasing and subcontract control
> Metallurgical and heat treatment departments
> Quality assurance department as a whole and especially including the following:
>
> The Calibration section
> Goods inwards inspection
> General inspection records
> Concession records
> Relevant subcontractor records appertaining to Company orders and covering as many of the foregoing areas as may be relevant

5.3. The responsible person in each of those departments and sections must ensure that the relevant records are compiled and maintained in a suitable manner and that they are readily accessible for examination by any authorized person, particularly including the quality assurance representative (QAR) of the MOD, or anyone nominated by the QAR.

5.4. The reports that are made as a result of audits on the performance of the various procedures in the Company quality system will also be suitably recorded and made available for examination in the same way.

5.5. As a whole the records will indicate the overall approach of the Company to the quality management program and also the use that is being made of those records for management control.

5.6. *Disposal of Records*

5.6.1. As reference may need to be made to any of the records covered by this procedure during the lifetime of the products concerned, the records that are covered by it may not be disposed of until and unless the written authority of the QAR has been obtained. However, if for any reason it is considered that it is necessary or desirable to dispose of records before any authorization from the QAR has been given, the QAR can be approached at any time with a disposal request.

5.6.2. Requests made to the QAR for disposal of records must contain full details of those proposed for disposal, the time period or other suitable indication of coverage, the reason for disposal, and the method proposed for the disposal.

5.6.3. If, and only if, the QAR agrees in writing, the designated records can be sent for disposal as agreed.

14
Quality Costs

The subject of quality costs is not one about which buyers are very familiar. Nevertheless, it is one of considerable importance for companies in general. There are, of course, considerable variations in the level of these costs from company to company. But as a general guide, it is usually considered that 10% of company sales income is a reasonable average for the quality costs for industry as a whole. For some companies the percentage will be less than this, although cases in which quality costs are as low as 5% or less will be few and far between. On the other hand, some companies have quality cost percentages in the range 10 to 15% and in a few instances, as high as 20% or even 25% (but see the postscript to this chapter). In 1978 a British government discussion document estimated that the annual quality costs (LOST COSTS) for British industry as a whole amount to about $20 billion.

For all practical purposes very little attention has been paid to costs and costing systems in the quality field until comparatively recently, and even that is not too much, although it must be said that there is now a British Standard, BS 6143, "Guide to the Determination and Use of Quality Related Costs." The government is pressing the benefits of quality cost improvement practices as part of its national quality campaign.

However, the attention that has gradually been turned toward quality costing has been largely because of a gradual recognition of the importance of costs in the quality field as a whole. It is probable that the work of the ITT Corporation begun in the mid-1960s on this subject first began to make companies realize the importance of reductions in quality costs. This realization is, however, being recognized by all too few companies. The fact is not yet fully appreciated that "quality costs" are a very significant part of a company's total expenditure and that intelligent planning can secure significant reductions in those costs. Although by no means the largest part of quality costs, costs associated with VQA operations, directly or indirectly, still represent a sizable chunk of money, with scope for worthwhile reductions. These cost reductions will be reflected immediately in a parallel improvement in the company's profit situation. It thus behoves VQA personnel to appreciate this cost situation and understand how they can influence it either way—for better or for worse.

The most obvious of the VQA associated costs are those prices a vendor charges the customer. These are well known. It is also well known that the vendor who usually gets the order is the one who quotes the lowest price. (Remember the thoughts of Colonel Glenn mentioned in Chapter 1.) Although this "principle" is beginning to fade away, it is not always appreciated that a customer very often, although they may not be aware of the fact, incurs significant additional costs with respect to a particular bought-out item when they use it—costs that, if known and added to the buying price, could place the price above that of competition.

The first of these VQA-related costs is that of the audit inspection service operated by BOQA. The reasons for this were explained in Chapter 6. There will always be a cost here, but it can vary quite a lot depending on whether the level of inspection is "reduced," "normal," or "tightened" (see MIL-STD-105D) and whether and how many shipments had to be rejected or, perhaps, 100% sorted. These all cost money to a greater or lesser extent. Also, it must not be thought that the cost of these rejected shipments is confined to the cost of inspection and return to the vendors. The latter may well be charged back, in any event. It should not take much thought (although few trouble to consider this factor) to appreciate that any cost incurred by the vendor will, sooner or later, and in some form or another, eventually be paid by customers in higher prices. What the VQA surveyor does when he or she helps to reduce the incidence of nonconforming product

QUALITY COSTS

delivered is to effect a cost reduction, which is usually well hidden although worthwhile. It could perhaps be regarded as "cost increase prevention."

It must be remembered that as the statistical basis for all sampling plans is very large numbers over a long period and average percentages of defects, it is thus almost inevitable that some unacceptable shipments will be accepted. If the BOQA operation does not succeed in preventing nonconforming products from being passed through to the production departments, other costs will be incurred. These will arise from the costs of stopped assembly lines, the cost of rework to defective assemblies, and the cost of possible scrap because defective items cannot be reclaimed (see the figures given later in the chapter). Whatever the cause of costs, they need to be established and apportioned to the cost charged by the vendor for the item in the first place. With this information, a buyer, with guidance from the VQA group, can consider on a factual basis the *true* cost to the company of buying from any particular vendor.

Identifying and establishing costs of this kind is not an easy task, and if it is to be done, a very careful plan will have to be prepared, probably with assistance from the company accountants. The record-keeping requirements for such a plan could be difficult for wholly written record systems. However, if computer systems are available, the difficulties will be very much less because of the high-speed capabilities of the computer. As one of the reasons for setting up a VQA group in the first place should have been to reduce the cost involved in using bought-out parts, it is suggested that even if it is not, or cannot, be done in the first instance, a long-term aim of the VQA operation should be just that: the identification and establishment of the amount of those usually neglected procurement costs.

ITT developed a comprehensive quality costs improvement plan which includes the following 12 principal steps:

1. Management commitment
2. Formation of the quality improvement team
3. Setting up the quality measurement program
4. Evaluation of the cost of quality
5. Initiation of the quality awareness program

6. Initiation of the corrective action program
7. Commencement of the defect prevention audit
8. Commencement of planning the zero-defect program
9. Commencement of the supervisor training program
10. Publicizing the zero-defect program
11. Initiation of the error cause removal program
12. Initiation of the task recognition program

Generally, quality costs are broken down into three principal categories:

1. *Appraisal.* These typically consist of all the costs associated directly with the inspection of production material, including personnel and equipment. They also include the costs associated with testing and statistical process control.
2. *Failure.* These typically include such items as scrap (internal and external), rectification, rework, reinspections, and warranty.
3. *Prevention.* Again, typically, these include design QA, quality planning, manual preparation, vendor assurance, auditing, and training. (VQA costs would fall into this category, as they are most certainly prevention.)

Of most obvious interest to buyers are those failure costs associated with the inspection of incoming material, the reinspections associated with the acceptance of nonconforming incoming material, and the costs of vendor quality assurance. There are, however, two other aspects of quality costs which are not particularly apparent, yet are of considerable importance to buyers. They are not included in any of appraisal, failure, or prevention categories. The first may simply be referred to as life costs. As for the second, I do not know of a recognized name for it, so I will call it simply the "hidden costs of buying." There are, therefore, four principal groups of quality costs of great interest to buyers: those associated with the incoming inspection both directly and indirectly, those associated with the operation of a formal system of vendor assurance, "product total life costs," and the "hidden costs of buying." They are discussed in turn in the following paragraphs.

Group 1. These costs are not likely to be directly susceptible to efforts on the part of buyers to reduce them, although they are well recognized and occur frequently. Generally, they arise when

QUALITY COSTS

either or both of two kinds of "accident" arises. In the first case the fact of an "accident" will not be readily recognized as such unless one looks at it from the point of view of the vendor. This is when a shipment is rejected at incoming inspection. Generally, three possibilities can result in such a rejection. In the first the shipment is simply rejected. No problem? Well, yes. The buyer will have to arrange for a replacement shipment, and usually on an urgent basis. This takes time and costs money. It is my belief that the cost of such an activity is seldom, if ever, determined. Usually, it is not even considered. But it is suggested that in the simplest cases the indirect, hidden costs could amount to the equivalent of at least 10 times the equivalent of the buyer's hourly pay as an absolute minimum.

In the case of the second possibility, for a variety of reasons, such as urgency or known nonavailability of immediate replacement, an engineering decision will have to be made to screen the shipment so that the nonconforming items can be segregated from the conforming ones. It may be known from the results of the sampling inspection that the proportion of nonconforming items, although more than that considered to be acceptable in terms of the specification, is not very much more. The good ones will be passed through for production use and the nonconforming ones returned to the vendor for eventual replacement. Here again, the cost to the buying department is seldom considered, just as in the case of the first possibility. Here it is suggested that the indirect, hidden costs will be at least an order of magnitude (at least 10 times) larger than that already suggested.

Possibility three is rather more complicated. In this case it will be known that the proportion of nonconforming items is relatively high and that, again, there are problems about nonavailability of replacements. There is also a time factor which prohibits the possibility of returning the items to the vendor for correction. So the buyer will have to enter into immediate and urgent negotiations with the vendor to negotiate two things: first permission for the customer to rework, or rectify, as felt appropriate, and second, a revised lower price for the substandard shipment and repayment to the customer of his or her incurred costs for reworking. Of course, such a rework cost has to be reasonable and be agreed to by the vendor—again an unknown cost of considerable magnitude.

The second type of case in the first group is when nonconforming items have escaped the net of incoming inspection and been passed through and assembled into finished product. This is a far more troublesome type of occurrence and one that could incur for the company very heavy costs indeed, depending on the stage of the overall manufacturing process at which the faults are found. Evaluating the costs in a situation of this type is, to say the least, not easy. However, a simple and very expressive way of representing the likely cost in cases such as these was devised some years ago in the United States. It is as follows.

Lost Costs: the cost of detecting an error

1. At the vendor — $1
2. At the customer's receiving inspection — $10
3. On the customer's assembly line — $100
4. At the customer's outgoing inspection — $1000
5. On the customer's end product in service — $10,000

Obviously, for many products the actual scale of lost costs will be less, even much less, than these. Equally, there can be little doubt but that in some cases the costs will be much more (e.g., in the Space Shuttle). So it should be clear that there is "big money" to be gained by preventing these cases from arising.

Group 2. The real value to the company of vendor assurance costs are much more difficult to evaluate and can only be so done over a fairly long period, both before the installation of such a system and after it has been put into operation. To a large extent the value of installing a vendor assurance system will be justified by the expression of a good deal of faith about the forecasts of the overall savings it will be expected to generate. Mostly, they will be in indirect, difficult to determine, overhead costs.

Group 3. In the United Kingdom the first widespread-investigation into the validity of aiming for "product total life costs" instead of the more usual "first costs" is believed to have been made by the Ministry of Defence. It resulted in an equally widespread beginning of a changeover in defense procurement to the "total life cost" concept, and the MOD was followed, although much more slowly, by industry.

The heart of the life cost concept is that when one considers the

QUALITY COSTS

cost of a project, all the costs from conception to eventual final disposal are taken into account. For our purposes some of these total costs are not very relevant, as they concern the costs incurred prior to the stage at which the buyer appears on the scene, that is, the design and basic development. But when the procurement stage is reached, the buyer can begin to consider the life cost. It is certain that many of the costs that arise after procurement will have been estimated by engineers. They will include such items as installation, maintenance, and servicing, including the cost of replacement parts. An important part of the investigative work to determine these costs will include estimation of the expected reliability of the product. Perhaps more than any other factor, the reliability factor will affect the overall life cost. But all those costs will have been evaluated and they will be ready and available for the buyer when he or she begins the task of "procuring."

What the buyer must do upon beginning his or her task of organizing the procurement of the item or product that is going to be ordered is many sided. The buyer must obtain from the prospective vendors their estimates of the various costs that will be incurred after the procurement task is completed. These must be compared on a summation basis with the quoted procurement cost that the vendor will charge the customer.

This comparison task is not easy. It must be done with care and the buyer will very definitely have to work closely with technical colleagues in the engineering (or equivalent) division of the company. Perhaps a simple analogy might be that of the person who wishes to buy, say, a twist drill for do-it-yourself activities. Usually, there are at least two qualities of twist drill available to buy. The first and cheaper version will probably be made of carbon steel. The second will be made of high-speed steel and will be more expensive. Consider the first choice. It will cost, say, $3.50, a price more attractive than that of the high-speed drill, say, $8.50. So at first sight one would feel inclined to buy the cheaper drill, as the price differential is large—2½ times. But consider the factors of "maintenance" and length of life. If this drill is going to be used for drilling mild steel, it is likely that we would eventually conclude that the high-speed steel drill would outlast the carbon steel drill by a factor of perhaps 6 or more. Looked at in this way, it is now very likely that the person would buy the high-speed steel drill, which would not blunt as quickly, would probably perform better, and would have a much longer life.

The foregoing is a very simple example of life-cycle costing. Although in most procurement projects the life cost analysis will deal with large sums of money, one must not forget that buyers can make many worthwhile, if small savings. The example of the drill just given is one in consumable stores. But the example discussed next demonstrates just how much costs can be reduced.

Group 4. The costs included in this grouping are perhaps the most difficult to appreciate but can readily be determined. The particular example that I like best is one with which I was closely involved not long ago. Only the bare facts will be given, but the reader will be able to fill in the complete picture without difficulty. The example is discussed more or less as it occurred.

One of my responsibilities before retirement was for a large gauge system. Orders for gauges were organized in the more or less traditional fashion. The gauge department prepared a requisition that was sent, with drawings, to the buyer who bought all gauges. There were about 20 or 30 gauge vendors on the buyer's "list," and he would usually pick four of them more or less at random and send an enquiry to each of the four with drawings.

In due course each of the chosen four would respond with a quotation for the gauge in question and the buyer would usually pick the one that had proposed the lowest price and send the order to that vendor. All very straightforward.

However, I was not very satisfied with the performance of many of those 20 or 30 gauge vendors and decided that the list should be cut to no more than a half dozen. After assessment visits were carried out at all of them, and with due consideration of the quality performance of each, the list was cut to six.

I then began to consider seriously the actual cost of the procurement of the gauges. An investigation was made and some rather surprising figures came to light. First, it was ascertained that the average range between the highest and lowest quotations in each usual group of four quotations was about $10; secondly, the average cost of the gauges ordered was about $20. Of course, the actual price of individual gauges ranged from about $7 to several hundreds of dollars. Obviously many more at the lower price range were ordered than the more expensive ones. So far, so good. The

QUALITY COSTS

next stage was to cost the various activities that occured during the buying cycle, including the provision of the necessary drawings that would go out with each individual inquiry. This was where the big surprise came. Taking all the indirect costs and their overhead allowances into account, the total cost to the buying department of the procurement of a single $20 gauge and saving $10 on the price was about $300.

After these figures were presented to the buying department, and thoroughly examined and discussed, a very significant change in the buying procedure was agreed to. No longer would requests go out for quotations from four prospective vendors. The buyer would select, in turn, one of the now greatly reduced list of gauge vendors and just send an order requesting that the price be confirmed when acknowledging. This new procedure was also welcomed by the gauge vendors themselves, for two quite significant reasons. First, they now had the assurance that each "enquiry"/order that they received was firm. Second, under the previous arrangement, they could be reasonably sure that less than 20% of quotations would result in an order. This meant that they, too, were incurring hidden procurement costs for the unaccepted quotations which were not shown in any account. In the long run, those LOST COSTS would be paid by customers in general. Although this example dealt specifically with gauges, it was observed that the same cost-reduction procedure could be operated with any item that required frequent purchases of relatively low value items.

Another fascinating example of hidden lost costs that would have been of considerable interest to the buying department of the company concerned (if anyone had told them about it) happened in New Zealand not long ago. Before describing the circumstances, a little background information will be helpful.

In most metalworking operations there is an inevitable, unavoidable, proportion of scrap. This proportion is usually small, although depending on the operation in question, it can also be quite high. A simple example is the case of washers being punched out of a strip of steel. Usually left is a web, a strip of steel with holes almost the full width of the strip, from which the washers were punched, plus a lot of small disks that were the washer centers. This waste material is generally known as process scrap and "everybody" knows that it is unavoidable, so when scrap notes are handed in by a manufacturing department claiming an allowance for "process scrap," it would never be questioned—but, back to the New Zealand example.

The company carried out a large amount of tube manipulation on tubes about 1 to 2 inches in diameter. The raw material was purchased from a local supplier in lengths varying from 12 to 14 feet, according to the purpose. They were cut into lengths varying from 12 to 30 inches. However, the first operation was to cut a square end at the beginning of each long tube before cutting it into the short lengths required. Normally, one would not expect more than, perhaps, ⅛ to ¼ of an inch to be lost as process scrap. In fact, the amount that had to be cut off was 4 inches, and the same happened at the other end of the tube as well—and all this because the tube ends were squashed.

To me, as an outsider, this strange waste stood out like the proverbial sore thumb, especially as I could not think of any reasonable explanation for those damaged tube ends which occurred on *every* tube. The writer asked questions and the story unfolded. Tubes like this had been received from the same supplier for a long time, the squashed ends being caused by the "cutting" method used by the supplier. No one had questioned this, although it meant that the company was paying for 7% of tube that it could never use and which it should never have received or paid for. My visit to this company was not long enough to permit a personal follow-through on this story, although I was assured that the supplier would be closely questioned.

If one cares to look for them, LOST COSTS can be found in a wide variety of hiding places—by no means all of which are directly associated with the procurement of productive material. But there are more than enough areas of lost costs that can be identified if one would only look for them, and could be-recovered to the considerable benefit of one's employer.

Postscript. In this chapter, it was also said that more often than not, LOST COSTS amounted to about 10 to 15% of the gross sales income of a company going up to perhaps 25% in extreme cases. In later paragraphs examples were given of cases in which some of these quite unsuspected costs were revealed. Readers will have noted that not all of them were strictly quality costs. However, without stretching the "quality" of words very far, one can say that those examples do relate to the quality of service being supplied by

QUALITY COSTS

or to a company—on its own account or when considering customers.

In 1986, a major campaign was launched in the United States called the Business Improvement campaign. The campaign was sponsored by the American Society for Quality Control and the American Management Association, among others. Included in the campaign were brief case studies of a number of companies that have made well-worthwhile cost recoveries on the general lines discussed in this chapter. These companies include a number of leading business, insurance, banking, and medical organizations, as well as departments of the federal government. But perhaps most interesting of all was the "punch line" of the campaign:

POOR QUALITY TAKES 35 CENTS OUT OF EVERY $1.00 OF GROSS SALES

Finally, it seems most appropriate to bring to the attention of readers an aphorism coined a good many years ago by J. M. Juran, the American quality management and business consultant of worldwide reknown. It is:

The GOLD in the MINE.

The gold in the mine is that 10, 15, 20, 25, or even 35 cents of LOST COSTS in each sales dollar.

Appendix
Standards

This appendix provides details about the various standards that have been mentioned in this book. They are listed in several ways, including the equivalence that many of them have with one another. Apart from that, they are listed by country and by the chapters in which each is mentioned.

The United Kingdom

BS 308.   Engineering Drawing Office Practice.

BS 5179.   The first BS, which described a range of three levels of quality system requirements for industry intended for use by commercial manufacturing companies. It was, generally, based on the Ministry of Defence (MOD) series of Defence Standards (Def. Stans.) in the series 05-21 to 05-29, which were issued in the early 1970's not long before BS 5179 itself.

BS 5750.   Essentially this is an improved version of BS 5179, which it superseded toward the end of the 1970s. It has six parts, of which the first three detail the requirements for the three levels of quality system requirements (Part 1 is the highest level). Parts 4 to 6 are guides to the implementation of Parts 1 to 3, respectively. It has had a considerable influence in other countries, some of which have more or less adopted it as their own national standard.

It is widely believed that BS 5750 is the base from which the new ISO 9000 series was developed. As a result of the issue of the ISO series in mid-1987, BS 5750 has had minor changes made to it to harmonize it with the 9000 series and has been reissued under dual numbering (i.e., BS 5750/ISO 9000 to 9004). BS 5750 is now in four parts, as follows. Part 0 is both a guide to companies setting up or developing quality systems and a guide to the practical implementation of the three levels of system requirements that appear in Parts 1 to 3, again with Part 1 as the highest of the three levels.

BS 6001. This is a set of attribute sampling inspection plans.

BS 6143. Guide to the determination and use of quality-related costs.

05-21 to 05-29. This is the series of U.K. Def. Stans. introduced in 1973 to describe for the first time in the United Kingdom a set of three levels of requirements for quality systems to which contractors to the Armed Forces would be expected to conform if they were to continue as defense contractors. Also, for the first time, full responsibility was placed on those contractors for the conformance of the product supplied to the MOD. This super-seded the original system, used since the formation of the Royal Air Force in 1919, under which the inspection departments of the MOD and its predecessors had themselves been responsible for the acceptance of that which they bought. It is understood that those Def. Stans. were based on the North Atlantic Treaty Organization (NATO) Quality Assurance Procedures (AQAP's) In turn, these are based on the American MIL-Q-9858A of 1963. Since 1985 these Def. Stans. have been replaced by the equivlanent AQAP's. 05-21 and 22 are, respectively, the system requirements for the highest of the three levels and a guide to its application. 05-24 and 25 serve the same purposes for the median level. 05-26 and 27 describe the measurement and calibration system requirements and their application, respectively. 05-29 describes the requirements for the lowest level. There is no guide for this.

Def. 131A. The military equivalent of BS 6001 attribute sampling plans, **DG 7A**; an early guide to the use of Def. 131A.

The NATO AQAP series. A small number of these, numbered 1, 2, 4, 5, 6, 7, and 9, are the procedures on which Def.Stans. 05-21

APPENDIX

to 05-29 were based. Since 1985 they have replaced those Def. Stans. in the MOD systems. They were based, in a much expanded manner, on MIL Q-9858A.

The United States

ANSI/ASQC Q1-1986. a standard that deals with the assessment of contractors.

ANSI Z1,4. A nonmilitary version of MIL-STD-105D.

MIL-Q-9858A. Originally published in 1959, this is a military standard for quality system requirements. It is understood to be the "father" of all the British military and civil, NATO, and ISO standards for contractors' quality systems that have been developed since then. The current issue is that of 1963.

MIL-STD-105D. Military set of attribute sampling inspection plans.

Australia

AS 1821 to 1823. Civil standards very similar to the British BS 5750.

Canada

Z 299. The civil standard which, for Canada, is the approximate equivalent of BS 5750 except that it has four levels of quality systems.

International

The ISO 9000 series. This recently promulgated (mid-1987) series of standards from the ISO consists of a set of five standards as follows; 9000 is a guide to the selection and use of 9001 to 9003. 9001 to 9003, themselves, consist of a set of three levels of quality system requirements which are broadly the same as the three levels of BS 5750. 9004 is intended to be used as a guide to companies setting up or developing quality systems as well as a guide to the implementation of the three levels themselves.

The chapters in which references to these standards are to be found are as follows:

Preface: B.S.* 5750; 05-21 series; MIL-Q-9858A; MIL-STD-105D: ISO (9000 series)
Chapter 1: Appendix 1: 05-21; 05-26
Chapter 2: None
Chapter 3: Appendix 1: 05-21
Chapter 4: BS 5750
Chapter 5: BS 5750; BS 6001; 05-21 series; Def. 131A; DG 7A; AQAP series; AS 1821 to 1823; MIL-SID-105D; ANSI/ASQC Q1-1986; ANSI Z1.4; Z 299; ISO series; (Appendix 4) BS 5750; AQAP 4; AQAP 6; (Appendix 5) ISO 9000 series; MIL-Q-9858A
Chapter 6: BS 5750; AQAP series; MIL-Q-9858A; (Appendix 1) 05-21; Def. 131A
Chapter 7: None
Chapter 8: None
Chapter 9: BS 5179; BS 5750; 05-21 series; ISO 9000 series; AQAP series; MIL-Q-9858A
Chapter 10: BS 308; 05-21
Chapter 11: None
Chapter 12: BS 5750
Chapter 13: Appendix 1; 05-21
Chapter 14: BS 6143; MIL-STD-105D

Standards That Have Near Equivalence

Systems for quality assurance. BS 5179; BS 5750; AQAP series; MIL-Q-9858A; ISO 9000 series; AS 1821 to 3. Z299

Attribute sampling inspection plans. Def. 131A; BS 6001; MIL-STD-105D; ANSI Z1.4.

By class groupings. U.K. Military; 05-21 series; Def. 131A; DG 7A.

NATO Military

AQAP series
U.S. military: MIL-Q-9858A; MIL-STD-105D
U.K. civil: BS 308; BS 5179; BS 5750; BS 6001; BS 6143
U.S. civil: ANSI/ASQC Q1-1986; ANSI Z1.4
Australian civil: AS 1821 to 1823
Canadian civil: Z 299

Index

Acceptable Quality Level, 99
Accounts department
  information loop, 96
Allied Quality Assurance
    Procedures, 139
AQAP
  from NATO, 49
Assessment
  exercise, 44
  method of, 54
  of procedures, 53
  records, grading, 177
  standards, differences in, 169
  team, 46
Assessor
  definition of, 168
  lead, 168
Audit
  grading system, 118
  inspection
    costs, 200
    planner, 99
  specification, 117, 151

Bought out parts

  nonconforming, 160, 161
Bought Out Quality Assurance, 93
  as audit function, 97, 98
  planner guide, 117
  sampling plans for, 149
  shipment documentation, 98
  vendor rating system, 97
  work areas, 104
British Standards
  308, 148
  5179, 140
  5750, 40, 140
    levels of, 51
    median requirements, 52
  6001, 48
    inspection sampling tables, 101
  6143, 199
Buyers
  activity
    advisory, 20
    consultative, 21
    familiarization, 21
  best decision, 3
  commercial considerations, 43
  confusion from specification, 19

215

INDEX

Buyers (cont'd)
  criticisms of, 1
  department, 3
  engineering, 1
  foundation of authority, 21
  information from vendors, 44
  new opportunities, 2
  order pack, 132
  restrictions, 15
  rewards for, 2
  specialist sections, 3
  traditional, 2
Buying
  cost reduction, 6
  department, 3
  problems, 4
  programs, 3

Case studies
  syndicate working on, 169
C A S T
  concept of, 21, 125, 129, 145
Chemical inspection
  of incoming material 105, 111
  laboratory facilities, 160
Commodity shortages, 3
Concession
  defined, 183
Costs
  of audit inspections, 200
  hidden, 202, 203
  increase prevention, 201
  of installation, 205
  lost, 203
  of maintenance, 205
  of replacement, 205
  of servicing, 205
  vendor estimates of, 205
Cubit
  variation, 17
Customer
  complaints, 24, 101
  drawing office number
    electrical, 38
    mechanical, 38
  reference catalogue number, 38
  requirements for vendor, 47
  satisfaction, 129

Defence
  procurement
    lower costs, 5
  standards
    series 05-21 to 05-29, 49, 139, 140
Department of Trade and Industry, 170
Design department
  basic quality responsibilities, 27
  executive arm, 160
  operational procedure, 26
Distributors, 43
Documentation system
  development of, 180

Economic indicators, 4

Forecasting, 4
Failure costs
  concept of, 4
First costs, 6

Glenn, Colonel, 5

Hill, Dr. David, 48

Incoming inspection
  and B O Q A, 160
  general procedures, 105
  problems, 131
  and V Q A, 160
Incoming materiel
  control of, 105, 113
Indicators, economic, 4
Information feedback loop
  buying and quality, 96
Inspection
  adjustments, 101

INDEX

Incoming inspection (*cont'd*)
  conditions clause, 39
  data, 101
  features, 99, 100
  incoming, 93
  mandatory, 121
  normal, 102
  planned, 15
  planner guide, 104
  of purchased parts, 93
  reduced, 102
  sampling, 48, 138
  tightened, 102
Inspectors
  instructions for, 151
Institute of Quality Assurance, 170
Instructions
  quality of, 176
Item acceptance specifications, 99
ITT
  quality
    cost improvement plan, 201
    cost reduction, 6

Laboratory facilities, 160
Lead
  assessor
    basic requirements, 170, 171
    certification board, 168, 170
    registered, 168
  problems, 4
Lost costs, 151
  containment of, 26

Market behavior, 4
Material substitution, 4
Measurement methods
  direct, 160
Metallurgical inspection
  of incoming materiel, 105, 111
  laboratory facilities, 160
Micrometer
  measurement for inspection, 103

MIL-Q-9858, 139
Mil-S-105D, 48, 101
Ministry of Defence, 4, 169
  assessment method, 163
  standards for quality assessment systems, 165
Models, statistical
  development of, 4

National standards, 140
North Atlantic Treaty Organization, 139
  Standardization Agreements, 50
  standards for quality assessment systems, 165

Operational procedures
  concessions and production permits, 183, 189
  documentation
    for incoming inspection, 183, 191
    purchasing, 183, 194
  preparation of, 149
  records
    retention and disposal of, 183, 195
    subcontractor, 183, 189
  standard format, 149
Overhead charges, 144

Packaging
  special requirements, 40
Part inspection
  specification for, 151
Personnel recruitment
  internal, 180
Planners
  audit inspection, 99
  method of inspection, 103
  resource, 4
Pollution
  reduction in, 3
Portsmouth Management College, 169

Procurement
  cost of, 40
  quality aspects of, 40
Product
  planning, 4
  quality assurance, 160
Production
  permit
    defined, 183
  samples
    initial submission of, 164
Programming methods, 176
Project life, 4
Purchasing
  diamond, 20
  department guide, 104
  research, 4

Quality
  Assurance, Institute of, 170
  control system, 163
  investigative work, 44
  manual, 53
  purchasing responsibilities, 9
  rating, determination of, 164
  systems, national standards for, 44
Questionnaire, vendor, 44, 45

Records
  B.O.Q.A., 177
  computer, 104
  system options, 176
  written, 175
Resource planning, 3
Resources
  future use, 3
Royal Air Force, 139

Sample component examination
  center for, 160
Sampling
  inspection, 48
  plans, 121

statistical protection in, 102
Security
  in inspection areas, 95
Shipping
  special requirements for, 40
Shipments, 94, 95
Solomon's Temple, 17
Specifications
  for acceptance inspection, 25, 94, 104
  audit, 117, 151
  compatibility, 26
  confusion, 18
  customer, 17
  errors, 26
  example, 19
  interpretation of, 14
  irrelevant data in, 148
  item acceptance, 99
  language for, 14
  modifications in, 132
  in Old Testament, 17
  for part inspection, 151
  purpose of, 16
  quality problems resulting from, 14
  requirements, 133
  technical consequences, 14
  as tool for buyers, 20
  type
    written or pictorial, 148
  writers, 14, 149
Staff training
  problems, 177
Standard
  forms, typical listing of, 180
  procedures
    breaches in, 40
Standardized method of training, 169
Standards, laboratory, 160
Statistics
  methods
    guidance in, 160

INDEX

Statistics (cont'd)
  models
    developments of, 4
    sampling inspection, 138
Storage facilities
  quarantine and bonded, 149
Supplier
  confusion, 23
  engineering capabilities and prospects, 3
Support services
  specialized, 161
Surveillance visits, 164
Surveyor
  definition of, 168

Target quality indicator, 164
Trade and Industry, Department of, 170
Training
  participative method, 169
  program, in-house, 170

Vendor
  assessment of, 46, 164
  assessment visit, 49
  capabilities, 45
  conformance records of, 102
  evaluations, 43
  guidance from buyers, 19
  grading of, 117
  quality
    manual, 47, 53
    performance of, 164
  questionnaire, 44
  rating
    data, 104
    determination of, 123
    price differentials, 163
    quality, 142
    system, 117
  relations program, 134
  surveillance visits, 125
Vendor Quality Assurance
  and buyers, 162, 164
  costs, 144
  definition of, 137
  effectiveness of group, 143
  lab facilities, 160
  manual, 149
  procedures, 147
  record keeping, 149
  specifications, 147
  technical back-up for, 160
  and vendors, 161
Vendor quality rating
  current, 179
  running, 179
Vendor Quality Surveyors, 131
  areas of operation for, 143
  selection of, 167
Vernier caliper
  measurement for inspection, 103

Waste
  reduction in, 3
Weighting factor
  arbitrary, 122
  critical faults, 122
World standard
  ISO 9000, 49, 140